DATE DUE

Demco

The Rand McNally Library of Astronomical Atlases
for Amateur and Professional Observers
Series Editors Garry Hunt and Patrick Moore

Jupiter

**Garry Hunt
and Patrick Moore**

Foreword by Professor Archie E. Roy

Published in Association with
the Royal Astronomical Society

Rand McNally & Company
New York · Chicago · San Francisco

This edition published by
Rand McNally & Company

Third printing, 1985

ISBN 0-528-81542-3 (hardbound edition)
ISBN 0-528-88073-X (paperback edition)

Library of Congress Catalog Card No. 80-53883

Jupiter was edited and designed by
Mitchell Beazley Publishers,
Artists House, 14-15 Manette Street
London W1V 5LB

Editor: Gilead Cooper
Designer: Wolfgang Mezger
Editorial Assistant: Charlotte Kennedy
Picture Research: Elizabeth Ogilvie

Executive Editor: Lawrence Clarke
Art Manager: John Ridgeway
Production Manager: Barry Baker

Printed in the United States of America
by Rand McNally & Company

The units and notation used throughout this book are based on the Système International des unités (SI units), which is currently being introduced universally for scientific and educational purposes. There are seven "base" units in the system: the *meter* (m), the *kilogram* (kg), the *second* (s), the *ampere* (A), the *kelvin* (K), the *mole* (mol) and the *candelo* (cd). Other quantities are expressed in units derived from the base units; thus, for example, the unit of force the newton (N) is defined as the force required to give a mass of one kilogram an acceleration of one meter per second squared ($kg\,m\,s^{-2}$).

Some branches of science continue to adhere to a few of the older units, and in one case an editorial concession has had to be made to existing scientific usage: the S1 unit of magnetism, the tesla, has been dropped in favor of the more common unit, the gauss. One tesla is equal to 10,000 gauss.

For very large and very small numbers, "index notation" has been adopted, so that where appropriate numbers are written as powers of ten. For example, 1,000,000 may be written as 10^6, and 3,500,000 as 3.5×10^6. Numbers smaller than one are indicated by negative powers: thus 0.00035 is written as 3.5×10^{-4}. In addition a variety of prefixes is used to denote certain multiples of units (*see* Table 1). Table 2 gives the SI equivalents of common imperial units while Table 3 lists a selection of astronomical constants.

Table 1: SI prefixes

Factor	Name	**Prefix** Symbol
10^{18}	exa	E
10^{15}	peta	P
10^{12}	tera	T
10^{9}	giga	G
10^{6}	mega	M
10^{3}	kilo	k
10^{2}	hecto	h
10^{1}	deca	da
10^{-1}	deci	d
10^{-2}	centi	c
10^{-3}	milli	m
10^{-6}	micro	μ
10^{-9}	nano	n
10^{-12}	pico	p
10^{-15}	femto	f
10^{-18}	atto	a

Table 2: SI conversion factors

Length	
1 in	25.4 mm
1 mile	1.609344 km
Volume	
1 imperial gal	4.54609 cm^3
1 US gal	3.78533 liters
Velocity	
1 ft/s	0.3048 $m\,s^{-1}$
1 mile/h	0.44704 $m\,s^{-1}$
Mass	
1 lb	0.45359237 kg
Force	
1 pdl	0.138255 N
Energy (work, heat)	
1 cal	4.1868 J
Power	
1 hp	745.700 W
Temperature	
°C =	kelvins − 273.15
°F =	$\frac{9}{5}$ (°C) + 32

Table 3: Astronomical and physical constants

Astronomical unit (A.U.)	1.49597870×10^8 km
Light-year (l.y.)	9.4607×10^{12} km = 63,240 A.U. = 0.306660 pc
Parsec (p.c.)	30.857×10^{12} km = 206,265 A.U. = 3.2616 l.y.
Length of the year	
Tropical (equinox to equinox)	$365^d.24219$
Sidereal (fixed star to fixed star)	365.25636
Anomalistic (apse to apse)	365.25964
Eclipse (Moon's node to Moon's node)	346.62003
Length of the month	
Tropical (equinox to equinox)	$27^d.32158$
Sidereal (fixed star to fixed star)	27.32166
Anomalistic (apse to apse)	27.55455
Draconic (node to node)	27.21222
Synodic (New Moon to New Moon)	29.53059
Length of day	
Mean solar day	$24^h03^m56^s.555 = 1^d.00273791$ mean solar time
Mean sidereal day	$23^h56^m04^s.091 = 0^d.99726957$ mean solar time
Earth's sidereal rotation	$23^h56^m04^s.099 = 0^d.99726966$ mean solar time
Speed of light in vacuo (c)	$2.99792458 \times 10^5\ km\,s^{-1}$
Constant of gravitation	$6.672 \times 10^{-11}\ kg^{-1}\,m^3\,s^{-2}$
Charge on the electron (e)	= 1.602 coulomb
Planck's constant (h)	$= 6.624 \times 10^{-34}$ J s
Solar radiation	
Solar constant	$1.39 \times 10^3\ J\,m^{-2}\,s^{-1}$
Radiation emitted	$390 \times 10^{26}\ J\,s^{-1}$
Visual absolute magnitude (M_v)	+4.79
Effective temperature	5,800 K

Contents

Foreword

Astronomy is the oldest of the sciences, born of the fact that man evolved on a planet from which he could see the sky. For 50,000 years he has had intelligence enough to study the heavens and attempt to understand what he saw by day and by night. And there is no doubt that what he saw and deduced from his observations has had extraordinary effects upon his life in many lands.

If our civilization had developed in the way it has on a planet whose skies were eternally cloud-covered (which is doubtful), we would have believed, up to some forty years ago, that the Earth *was* the universe. Only with the advent of large radar dishes, high-flying aircraft and rockets would the shattering fact have emerged that above the opaque cloud-layer lay a seemingly boundless universe.

Modern western civilization had been greatly influenced by Copernicus, Kepler, Galileo and Newton, all watchers of the skies. Our belief in a rational universe capable of being understood and our scientific and technological civilization spring from the cyclic behaviour of Sun, Moon, planets and stars and Newton's ability to explain so much of that behaviour by his law of gravitation and his three laws of motion. Timekeeping, navigation, geodesy, dynamics, religious and philosophical systems, cosmology and relativity and many other activities and interests of man have been directly affected by our study of the heavens.

There have been three astronomical revolutions. The first—the serious, naked-eye study of the heavens—lasted a long time—at least five millennia—and ended when Galileo began his systematic telescopic study of the sky in AD 1610. That second revolution, in which the camera and the spectroscope played their part, was brought to a climax by the enormous amount of information gathered by telescopes such as the 200-inch Hale telescope at Mt Palomar. On October 4, 1957, with the orbiting of Sputnik I, the third astronomical revolution began. Not only can we now place instruments in orbit above the Earth's atmosphere, obtaining access to the entire electromagnetic spectrum, but we can send spacecraft such as the Mariners, Pioneers and Voyagers to other planets in our Solar System. The flood of astronomical information has become a torrent, sweeping away many of our former ideas about the universe.

The time therefore seemed ripe for a series of atlases designed to take stock of this flood of new information and the new understanding it has brought us of the nature of the universe. Each atlas in the series has been written by an author chosen by Mitchell Beazley Publishers so that the text will provide the most up-to-date assessment of the celestial body studied, together with explanatory diagrams and the most modern pictures. Each author's text has then been carefully checked and authenticated by an acknowledged expert in the subject chosen by the Royal Astronomical Society's Education Committee. The final text of each book should therefore truly convey our present-day knowledge of the subject and remain a definitive work for many years to come.

Archie E. Roy
BSc, PhD, FRAS, FRSE, FBIS
Titular Professor of Astronomy in the
University of Glasgow
Chairman of the Education Committee of the
Royal Astronomical Society

Introduction

Of the five planets known since very early times, two stand out at once because of their brightness: Venus and Jupiter. Venus is much the brightest, but its movements showed that it was the closer of the two and it was therefore considered as the less important. Jupiter was regarded as the senior member of the planetary family, and it was named in honor of the chief god of the classical pantheon.

To astrologers, who charted its movements before the invention of the telescope, Jupiter was of special significance, with effects that were generally benign. Its movements were closely studied and accurate tables of its motion were drawn up. Jupiter is usually well placed for observation for several months in each year, so that it is a familiar object indeed.

Inevitably it was one of the first planets to be studied telescopically, and surface details became clearly discernible for the first time. The telescope also led to the discovery (made almost simultaneously by Galileo and Marius) of the four brightest of Jupiter's satellites, now known as Io, Europa, Ganymede and Callisto, after four of Jupiter's lovers in classical literature.

But the real nature of Jupiter remained a puzzle, and until our own century it was widely believed to be a kind of junior sun, sending out enough heat to warm its system of satellites.

Yet this is not the case. Jupiter is not a star; it is a planet, and the temperature even at its core is far too low to trigger off nuclear reactions. Today it is believed that while the outer layers are undoubtedly gaseous, most of the vast globe is liquid, with only a small solid core. Jupiter is quite unlike the Earth or any of the other inner planets, and it is of tremendous importance. It has even been said, rather jocularly, that the Solar System is made up of the Sun, Jupiter and assorted debris!

During the past ten years four space-probes have bypassed Jupiter: two Pioneers and two Voyagers. They have increased our knowledge of the planet dramatically, and the information which they provided forms the basis of this book; the state of scientific knowledge before the space-probes is also described (*see* pages 10–11). The spectacular photographs sent back by the space-probes have revealed volcanic activity on the satellite Io, the existence of a faint ring round Jupiter itself, and the presence of three previously unknown satellites—to name only some of the surprising discoveries. For the first time it has been possible to make detailed maps of the Galilean satellites.

Nevertheless, there is still a great deal to be learned, and observation with Earth-based telescopes is still as important as ever. What has to be remembered, above all, is that the surface of Jupiter is changing constantly. Features appear and vanish; they change in form, in prominence and in color, so that one can never forecast exactly what will happen next. Amateur observers have carried out outstanding work in this field, and it is probably fair to say that in the pre-Space Age most of our knowledge of the behavior of the surface features was due to amateur research. Though no Earth-based telescope can show anything like the amount of detail provided by the space-probe pictures, it is worth commenting that at the present time there are no probe pictures being received; the Pioneers and Voyagers have gone on their way, and will never return, so that in the foreseeable future they will leave the Solar System forever. The next Jupiter pass is not scheduled before the mid-1980s, so that for the moment we have again to rely upon what may be termed "old-fashioned" visual and photographic observation.

Because Jupiter is so large, its main features—and even some of the minor ones—are accessible to amateurs using telescopes of modest aperture, say 15 to 30 cm. Systematic observations are of great value to the theoreticians, and it is pleasing to find they are still being produced in great quantity.

Jupiter is a magnificent world, and one of surpassing interest. We can never hope to send astronauts there, but we can continue to do our best to find out what the Giant Planet is really like.

1

2

3

1. The Planet Jupiter
Photograph by G. P. Kuiper, Catalina Observatory, Texas.

2. Symbol of Jupiter

3. The god Jupiter

Jupiter in the Solar System

Jupiter, the largest planet of the Solar System, is fifth in order of distance from the Sun. It is more than 1,330 times greater in volume than the Earth, and 318 times greater in mass. Indeed its mass accounts for more than two-thirds of the total mass of all the planets combined. It has 16 known satellites (see pages 50–84), four of which are themselves the size of small planets. The ancients aptly named it in honor of the ruler of Olympus.

Jupiter orbits the Sun at a mean distance of 5.203 astronomical units, equal to approximately 778,360,000 km. The orbit is not, however, circular, and Jupiter's distance from the Sun ranges from a maximum of 815,700,000 km (at "aphelion") to a minimum of 740,900,000 km (at "perihelion"). It takes 4,332.59 Earth days for Jupiter to complete one orbit (about 11.86 Earth years), but like the other giant planets of the Solar System, Saturn, Uranus and Neptune, Jupiter spins very rapidly about its own axis: the Jovian "day" is less than 10 hr long, so that there are approximately 10,600 Jovian "days" in one Jovian "year". This rapid rotation causes the planet to bulge noticeably at the equator, producing an ellipsoidal shape which is described as "oblate". Thus Jupiter's equatorial radius of 71,400 ± 100 km is substantially greater than its polar radius of 66,550 ± 100 km.

Although Jupiter is extremely massive (1.901×10^{27} kg), its mean density is only 1,330 kg m^{-3}, as compared to the Earth's 5,517 kg m^{-3}. The large size and low density indicate that Jupiter is composed principally of the lighter elements, in particular hydrogen and helium; these are present mainly in the form of gas and liquid. Every visible feature on Jupiter is, in fact, a cloud, and it is not known for certain whether Jupiter has any solid core at all.

Brightness

Jupiter, although extremely bright, is not actually the brightest of the planets as seen from Earth. Its maximum magnitude is −2.6, which is much less brilliant than Venus at its maximum of −4.4. On the other hand, Venus is closer to the Sun than is the Earth, and can therefore never be observed throughout the night, whereas Jupiter is favorably placed for observation for several months in every year. Mars can also outshine Jupiter at times, but the close approaches of Mars are very occasional. Even at its faintest, Jupiter shines far more brightly than Sirius, the brightest of the stars.

Occasionally Jupiter may pass in front of a star and hide or "occult" it. Similarly, Jupiter itself may be occulted by the Moon or by one of the nearer planets, but mutual occultations of planets are very rare indeed. The next occultation of Jupiter by a planet (in fact, Venus) will be on 22 November 2065. The last really "near miss" was in May 1955, when Jupiter was in the same low-power telescopic field as Uranus, while on 6 February 1892 Jupiter and Venus were separated by only 40 seconds of arc.

Oppositions

A "superior" planet such as Jupiter (one whose orbit lies outside that of the Earth) is best placed for observation when its position in the sky as seen from Earth is directly opposite that of the Sun. It is then said to be "in opposition". Jupiter comes to opposition once in about every 13 months, when the Earth has completed just over one orbit since the previous opposition and has caught up with the distance travelled by Jupiter during the terrestrial year. The interval between successive oppositions is known as the synodic period; Jupiter's mean synodic period is 398.9 days. Because Jupiter is so far from the Sun, its opposition magnitude does not vary greatly, unlike that of Mars. Its apparent diameter is also much greater than that of Mars, with a mean value of 46.86 sec of arc (for the equatorial diameter). As a result, surface details are considerably easier to see on Jupiter than they are on Mars. Moreover, even at its furthest from the Earth, Jupiter's apparent diameter is not less than two-thirds of the mean value, and its disc never appears too small for useful observations to be made.

1

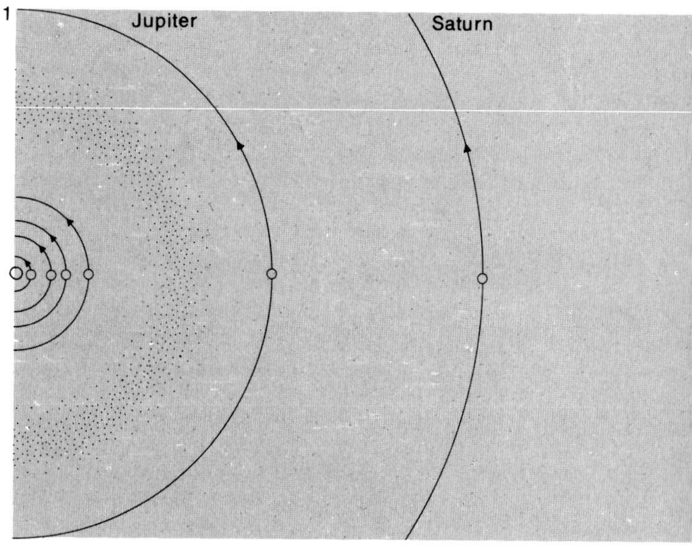

2

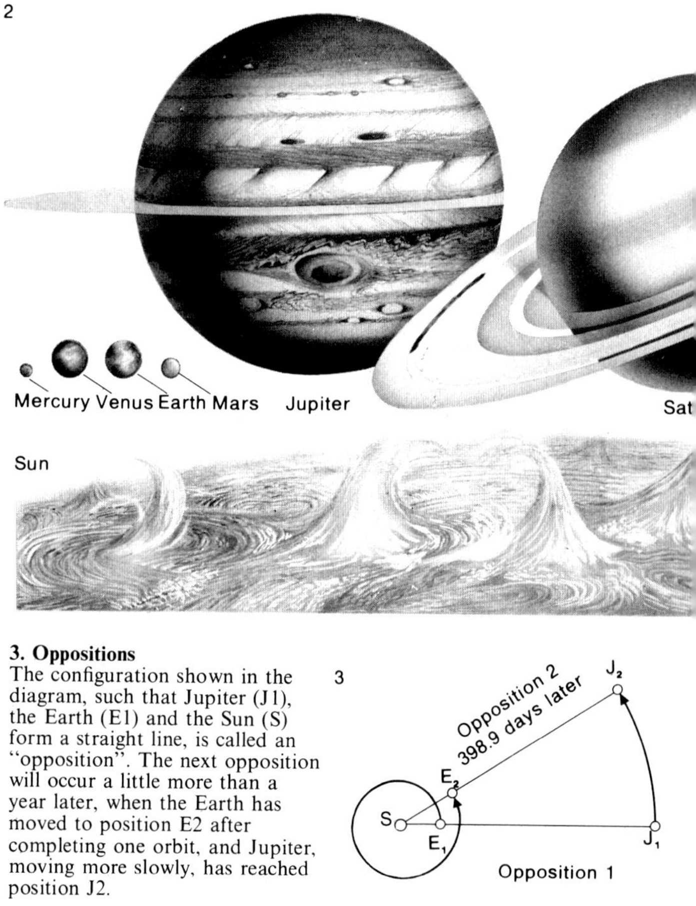

3. Oppositions
The configuration shown in the diagram, such that Jupiter (J1), the Earth (E1) and the Sun (S) form a straight line, is called an "opposition". The next opposition will occur a little more than a year later, when the Earth has moved to position E2 after completing one orbit, and Jupiter, moving more slowly, has reached position J2.

3

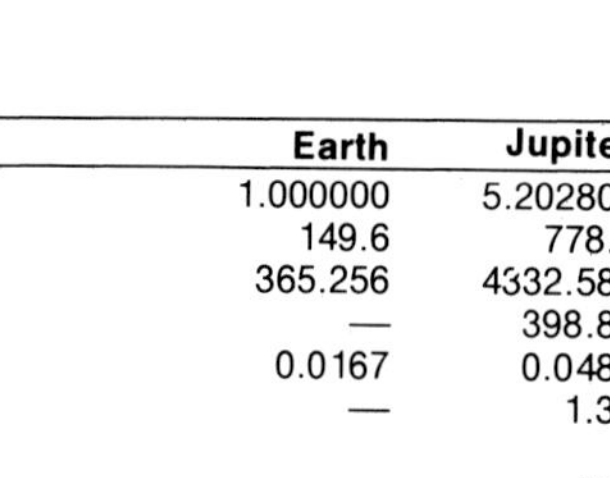

Mean orbital elements of Jupiter and the Earth

Orbital elements	Earth	Jupiter
Mean distance (astron. units)	1.000000	5.202803
(millions of km)	149.6	778.3
Sidereal period (days)	365.256	4332.589
Synodic period (days)	—	398.88
Eccentricity	0.0167	0.0485
Inclin. to the ecliptic (degrees)	—	1.30
Mean longitude of:		
Ascending node (degrees)	—	100
perihelion (degrees)	103	14

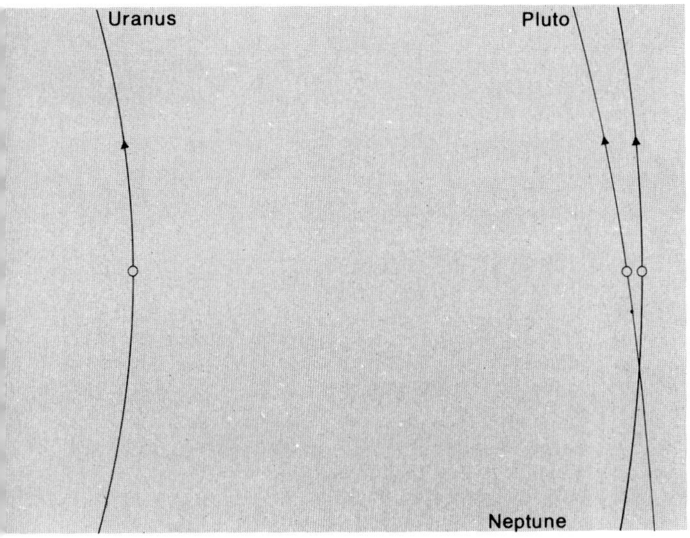

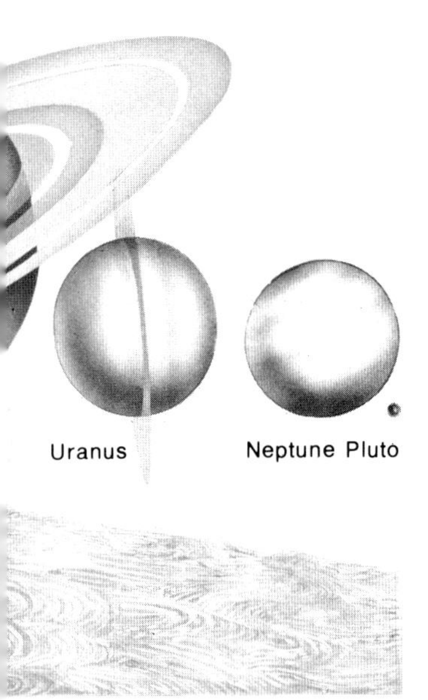

1. The Solar System
Nine major planets and many minor bodies orbit around the Sun. Jupiter is the fifth planet from the Sun; it is the nearest of the giant planets whose orbits lie beyond the asteroid belt, shown as a speckled region in the diagram. The inner planets, Mercury, Venus, Earth and Mars, are also shown within the darker shaded area. Pluto is shown within the orbit of Neptune, where it currently lies and where it will remain for approximately the next 20 years. Jupiter's orbit, like those of all the planets, is an ellipse, with the Sun at one focus. Its mean distance from the Sun is about 778,360,000 km, more than five times that of the Earth.

2. The scale of the planets
Jupiter, the largest of the planets, is drawn to scale in this diagram showing the relative size of the members of the Solar System. Part of the limb of the Sun is included for comparison.

Spectra

Jupiter's atmosphere is known to be very different in composition from that of the Earth. This information is derived largely from Earth-based studies of the spectrum of radiation received from Jupiter, and the technique also makes it possible to compare Jupiter's atmosphere with those of other planets, such as Mars and Earth (*see* diagram 4).

Atoms and molecules in a planet's atmosphere may emit or absorb radiation (depending on their temperature), and every substance emits and absorbs at certain characteristic wavelengths, just as a taut string will resonate only at given frequencies. The "resonant" wavelengths that characterize individual chemicals are accurately known from laboratory experiments. Thus the electromagnetic radiation received from a planet, both in the visible and the non-visible part of the spectrum, provides information about the constituents of the planet's atmosphere: their presence shows up as "lines" of absorption or emission at known wavelengths.

Early spectroscopic studies of Jupiter from the Earth confirmed that the atmosphere of Jupiter is composed mainly of hydrogen. Small amounts of ammonia and methane were also detected at an early stage and, as the spectral resolution of instruments improved, very small but important traces of other substances were found. The proportions of the substances detected in Jupiter's atmosphere (above the level of the cloud tops) are listed in the table below, with a comparative table for the Earth's atmosphere.

It is now known that helium is also present in the Jovian atmosphere in substantial proportions. However, the problem about observing spectra from the surface of the Earth is that the Earth's atmosphere also absorbs radiation, particularly in the ultraviolet part of the spectrum, and since the characteristic absorption lines of helium happen to occur at very short ultraviolet wavelengths, positive evidence of the presence of helium on Jupiter was not available before the Pioneer spacecraft measurement. With the addition of data from the Voyager mission (*see* pages 14–17) it has been established that the ratio of hydrogen to helium is close to that found on the Sun. Similarly, it has been shown that on Jupiter the ratio of hydrogen to carbon and nitrogen is also in near-solar proportions. This similarity in composition suggests the idea that Jupiter may be thought of as a "failed star". However, the analogy should not be taken too literally. Jupiter remains a planet, not a star, and although it does radiate a certain amount of its own energy (*see* pages 27 and 85), its mass would need to be at least 100 times greater for the thermonuclear reactions that produce the energy of a star to take place.

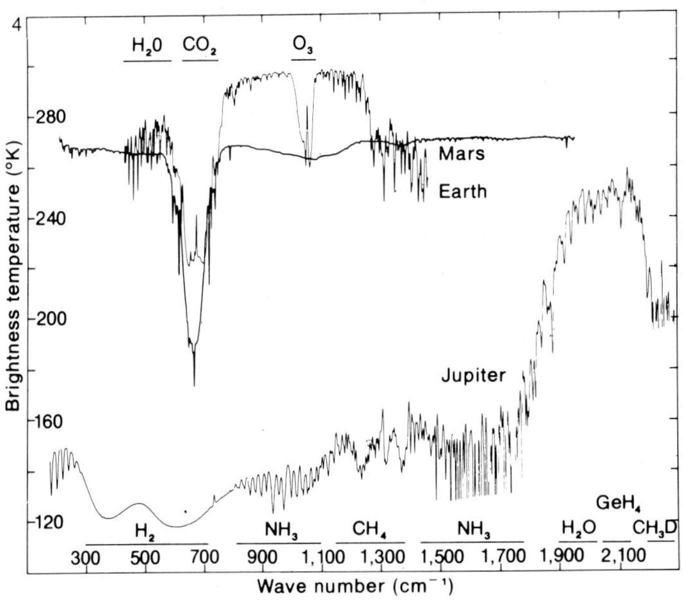

4. Spectra
The principal constituents of a planet's atmosphere can be identified by the way in which they emit and absorb radiation. This graph shows Jupiter's spectrum; the spectra of Mars and the Earth are also given for comparison. The graph plots the brightness temperature of the radiation against its wave number. Prominent spectral features indicating the presence of certain important molecules are marked.

Composition of Jupiter's atmosphere above cloud tops

	% volume
Hydrogen	≈ 90
HD	$\approx 1.8 \times 10^{-3}$
Helium	≈ 4.5
Methane	$\approx 7 \times 10^{-2}$
Deuterated methane	$\approx 3 \times 10^{-5}$
Ammonia	$\approx 2 \times 10^{-2}$
Ethane	$\approx 10^{-2}$
Acetylene	$\approx 10^{-2}$
Water vapor	$\approx 1 \times 10^{-4}$
Phosphine	$\approx 10^{-6}$
Carbon monoxide	$\approx 10^{-7}$
Germanium Tetrahydride	$\approx 10^{-7}$

Composition of the Earth's atmosphere

	% volume
Nitrogen N_2	76.084
Oxygen O_2	20.946
Argon A	0.934
Carbon dioxide CO_2	0.031
Neon Ne	1.82×10^{-3}
Helium He	5.24×10^{-4}
Methane CH_4	1.5×10^{-4}
Krypton Kr	1.14×10^{-4}
Hydrogen H_2	5×10^{-5}
Nitrous oxide N_2O	3×10^{-5}
Carbon monoxide CO	10^{-5}
Xenon Xe	8.7×10^{-5}
Ozone O_3	up to 10^{-5}
Water (average)	up to 1

The Nomenclature

Through the telescope Jupiter appears as a flattened yellowish disc crossed by dark streaks which have always been known as the "cloud belts". The bright bands are known as "zones". The surface features are always changing in detail, but several of the belts and zones are relatively stable, changing only slightly in latitude, while others periodically disappear for a while. The more permanent of Jupiter's surface features have been given a standard nomenclature, which is described in detail opposite. The abbreviations of these names are also standardized. The latitudes of the belts and zones as shown in the illustration on this page represent the mean positions of these features, measured from photographs taken over several years, rather than their appearance at a particular instant of time.

In the past it has been the normal practice to reproduce photographs and drawings of Jupiter with the south at the top, the way it appears to an observer in the northern hemisphere of the Earth using an astronomical telescope which gives an inverted image. Since the advent of space exploration, however, there has been a change in the official attitude: the more recent tendency, followed throughout this book, is to show north at the top of the picture or map.

Similar confusion can arise over the terms "east" and "west". In discussions of Jupiter, ambiguity is normally avoided by referring instead to the "preceding limb" (where features disappear as the planet rotates about its axis) or to the "following limb" (where they first appear). Thus with north at the top of the picture, the preceding limb is to the right and the following limb to the left.

Rotation

As a result of its predominantly fluid composition, Jupiter does not spin in the same way as a solid body. Instead, features at different latitudes on the planet rotate at different speeds, with those in the equatorial regions rotating the most rapidly. The movement of the visible surface of Jupiter is extremely complex, and for convenience the planet has been divided into two main regions: a separate rotational period has been adopted for each based on careful observation over a period of many years, and these two rates provide a fairly accurate guide to the relative movement of different features against which longitude can be measured.

The two rotational periods are known as Systems I and II. System I applies to the equatorial regions, bounded by about 9° of latitude on either side of the equator, and has a value of 9 hr 50 min 30.003 sec. This is equivalent to 877°.900 rotation in one Earth day. Within this region, however, there is a band of exceptionally rapid rotation known as the Great Equatorial Current. System II applies to everything outside $\pm 9°$ latitude, and is about 5 min longer than System I, since the polar regions move more slowly. Its exact value is taken as 9 hr 55 min 40.632 sec, or 870°.270 per Earth day, although once again this is an average over a range of values. At very high latitudes it is extremely difficult to determine the rate of movement accurately, because of the lack of clearly visible detail in these regions.

Tables are available giving the agreed longitudes of System I and System II at regular intervals. In order to determine the longitude of a particular feature, an observer would first decide whether it lay in the region covered by System I or System II, and then note the time at which the feature "transits" or crosses Jupiter's central meridian (the line joining the north and south poles). The difference between the time of the transit of the feature and the time at which longitude 0° of the relevant System last crossed the meridian may then be converted into degrees of longitude, making due allowance for the rate of rotation of the planet. (In practice the actual procedure is to convert the time of transit into degrees and add it to the longitude of the central meridian at time 0 hr—*see* page 86.)

There is, finally, a third system of rotation, known as System III, which refers to the source of certain "radio" emissions from Jupiter (*see* page 24).

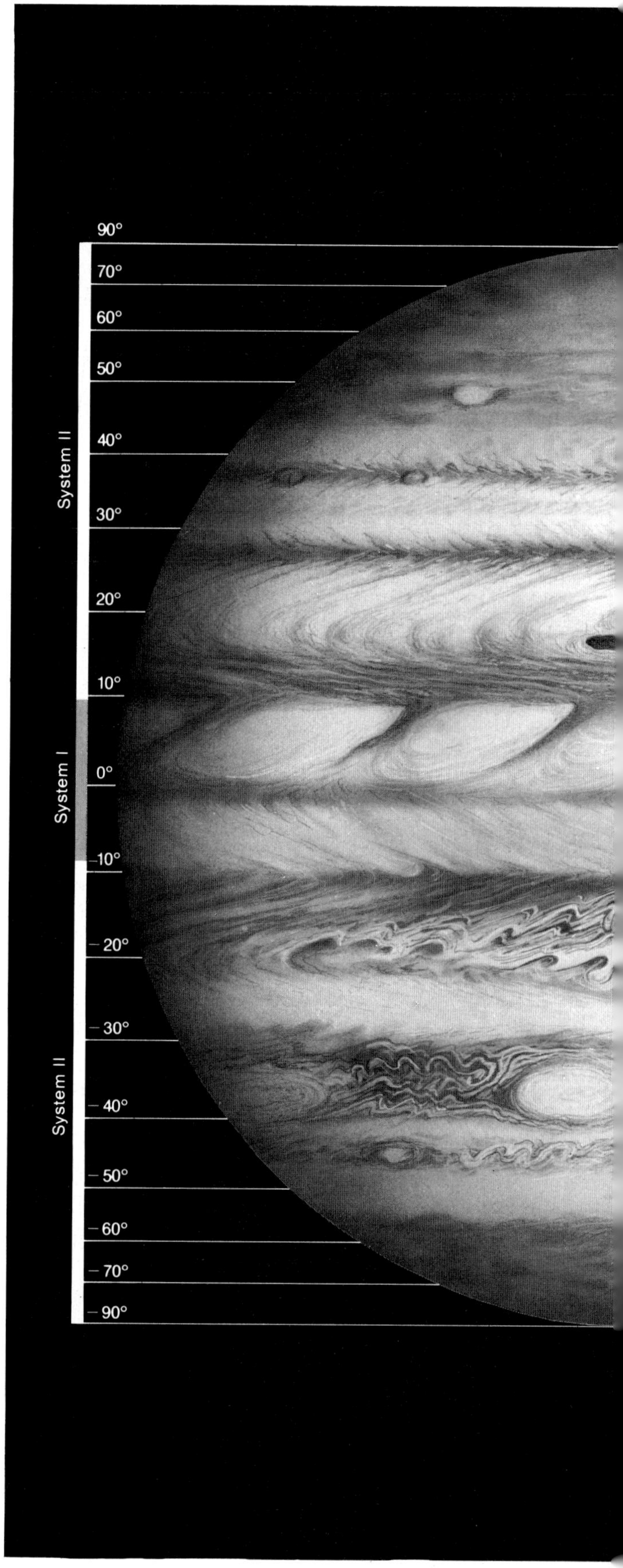

North Polar Region (NPR) Lat. +90° to +55° approx.
Usually dusky in appearance and variable in extent. The whole region is often featureless, though there are occasional white spots. The "North Polar Current" has a mean period of approximately 9 hr 55 min 42 sec.

North North North Temperate Belt (NNNTB) Mean Lat. +45°
An ephemeral feature often indistinguishable from the NPR; at other times reasonably distinct.

North North Temperate Zone (NNTZ) Mean Lat. +41°
Often hard to distinguish from the overall polar duskiness.

North North Temperate Belt (NNTB) Mean Lat. +37°
Occasionally prominent, sometimes fading altogether as in 1924.

North Temperate Zone (NTZ) Mean Lat. +33°
Very variable, both in width and brightness.

North Temperate Belt (NTB) Mean Lat. +31° to +24°
Almost always visible, with a maximum extent of about 8° latitude. Outbreaks occur in and near it; dark spots at its southern edge are not uncommon.

North Tropical Zone (NTrZ) Mean Lat. +24° to +20°
At times very bright. The "North Tropical Current", which overlaps with the North Equatorial Belt, has a mean period of approximately 9 hr 55 min 20 sec.

North Equatorial Belt (NEB) Mean Lat. +20° to +7°
The most prominent of all the Jovian belts. This region is extremely active and shows a large amount of detail, such as dark projections from the southern edge or white spots and rifts in the middle.

Equatorial Zone (EZ) Mean Lat. +7° to −7°
Covering about one-eighth of the entire surface of Jupiter, the EZ exhibits much visible detail. At the time of the Voyager encounter the northern component was dominated by 13 plume-like features. From Earth the whole zone has been found to abound with white ovals and streaks, and with wisps and festoons extending into it from the belts on either side.

Equatorial Band (EB) Mean Lat. −0.4°
At times the EZ appears divided into two components by a narrow belt, the EB, at or near to the Jovian equator. It may be doubted, however, whether this feature should be regarded as a true belt.

South Equatorial Belt (SEB) Mean Lat. −7° to −21°
The most variable of the Jovian belts. It is often broader than the NEB and is generally divided into two components by a region which at times appears almost as an intermediate zone. The southern component contains a "bay" known as the "Red Spot Hollow" (RSH).

South Tropical Zone (STrZ) Mean Lat. −21° to −26°
Contains the famous "Great Red Spot". The STrZ was the site of the long-lived "South Tropical Disturbance" and, in 1940–41, certain "oscillating spots".

Great Red Spot (GRS) Mean Lat. −22°
Although there are other spots on Jupiter's surface, both red and white, the GRS is much the most prominent. It rotates in an anticlockwise direction, and at present measures 26,200 km in length and 13,800 km in width: its surface area is thus comparable with that of the Earth. It may or may not be a permanent feature.

South Temperate Belt (STB) Mean Lat. −26° to −34°
This belt has never been known to disappear, although it is very variable in width and intensity; at times it appears double.

South Temperate Zone (STZ) Mean Lat. −38°
Often wide; may be extremely bright. Spots are common, particularly close to the edge of the STB.

South South Temperate Belt (SSTB) Mean Lat. −44°
Variable, with occasional small white spots.

South South South Temperate Zone (SSSTZ) Mean Lat. −50°

South South South Temperate Belt (SSSTB) Mean Lat. −56°

South Polar Region (SPR) Lat. −58° to −90° approx.
Like the NPR very variable in extent, sometimes reaching as far as the SSTB.

Early Observations

Before the age of space exploration most of the detailed information about changes in Jupiter's appearance came from amateur observers. The Jupiter Section of the British Astronomical Association (BAA), founded in 1890, has been particularly active. A summary of observational data up until about the middle of the present century is contained in the classic book *The Planet Jupiter*, by B. M. Peek.

The Great Red Spot

The most famous feature on the surface of Jupiter, the Great Red Spot, is known to have persisted for more than three centuries. Although it has disappeared at times, it has always returned, and for the past hundred years its behavior has been monitored almost continuously. A feature which may have been the Great Red Spot was seen by Robert Hooke in 1664, and the *Philosophical Transactions of the Royal Society* (Vol. 1., No. 3. 1664) contains the following account:

"... the ingenious Dr. Hooke did some months since intimate to a friend of his that he had, with an excellent 12 feet telescope, observed, some days before he spoke of it (viz. on 1664 May 9 os) about nine o'clock at night, a spot in the largest of the three observed belts of *Jupiter*; and that, observing it from time to time, he found that, within two hours after, the said spot had moved east to west about half of the diameter. It is situated in the northern part of the southern belt. Its diameter is one-tenth of *Jupiter*; its centre, when nearest, is distant from that of Jupiter about one-third of the semi-diameter of the planet ..."

The Italian astronomer Giovanni Cassini also reported observations in 1665, and the Spot continued to be observed intermittently until 1713. The next known record is a drawing made by Heinrich Samuel Schwabe, a German apothecary, in 1831. In 1857 it was drawn by William Rutter Dawes, an English clergyman, and in 1870 by Alfred M. Mayer, an astronomer at Lehigh University in Pennsylvania. Drawings made by the fourth Earl of Rosse in the early 1870s also show it unmistakably. It leapt into prominence in 1878, when observers described it as oval, striking and brick-red. Until 1882 it dominated the whole Jovian scene whenever it lay on the hemisphere turned towards Earth. It then faded, and observers began to fear that it would vanish permanently; but it revived in 1891, and since then it has been on view more often than not. There have been times, however, when it has been difficult to locate, such as in 1928–29, 1938 and 1977.

The size and position of the Great Red Spot also vary with time, as do those of the white ovals. It was at its largest in the 1880s, when it measured about 40,000 km by 14,000 km, nearly twice its present length. Although its latitude has remained roughly constant at 22 S, its longitude has drifted considerably, and over the past century the total change has amounted to more than 1,200°.

The South Tropical Disturbance

In 1900 some dark humps were observed extending from the South Equatorial Belts into the South Tropical Zone. They developed until they formed a shaded part of the whole South Tropical Zone, and the feature became known as the "South Tropical Disturbance" (STD). Initially its rotational period was 9 hr 55 min 20 sec, and since the period of the Great Red Spot at the time was 9 hr 55 min 41 sec it was evident that the Disturbance would eventually catch up with the Spot. The first conjunction occurred in June 1902. Surprisingly, the Disturbance did not overlap the Spot, but instead appeared to "leap" past it, reappearing on the far side; the entire event lasted less than three months. It was repeated regularly until the Disturbance faded gradually after 1935, and the last definite observation of it was in 1940. The records of these interactions have recently been interpreted in terms of the behavior of "solitons" (*see* page 31).

1. Drawings by Rev. T. E. R. Phillips
This selection of drawings shows Jupiter's changing appearance as recorded by one of the best-known amateur observers of the planet, and covers the years 1927 to 1929. Phillips was, for a while, director of the Jupiter Section of the British Astronomical Association, and contributed to the accurate measurement of the drift in longitude of various features.

1

26 July 1927

10 October 1927

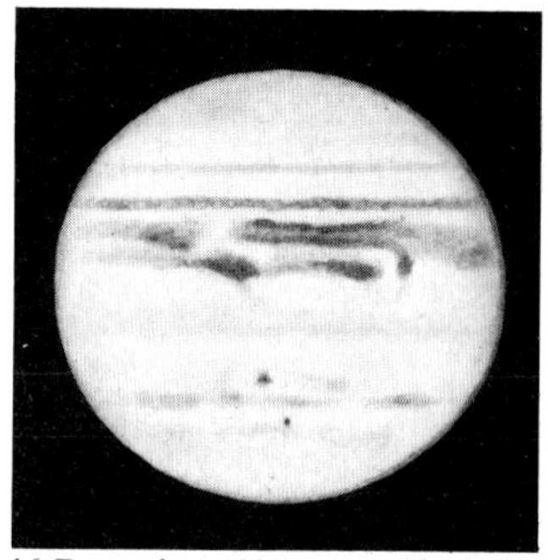

16 December 1927

11 September 1928

2 October 1928

2. Giovanni Cassini (1625–1712)
One of the early observers of the GRS, Cassini also measured Jupiter's rotational period.

2

The South Tropical Zone Circulating Current

Before the disappearance of the South Tropical Disturbance, some interactions were observed which gave rise to the name "Circulating Current" given to the prominent current of the South Tropical Zone. The South Tropical Circulating Current has two components flowing on either edge of the South Tropical Zone: the northern component flows in an easterly direction at approximately latitude 21°S, while the southern component flows in the opposite direction at approximately 26°S. Both components have a zonal velocity of about 50–60 m s^{-1}. Spots moving westward in the northern component were observed to move suddenly southward and join the southern component of the current when they came into contact with the preceding (eastern) edge of the South Tropical Disturbance. Thus the spots would approach the Disturbance, cross the South Tropical Zone and then return in the direction from which they came. Next, they would move towards the Great Red Spot from the west along the southern component with a relative velocity of 60 m s^{-1}, but would disappear as they came within 10,000 km of the western edge of the Spot.

Until 1965 no spot in the westerly component of the Circulating Current had ever been seen to survive an encounter with the Great Red Spot. In 1965, however, a small double spot was observed approaching the Great Red Spot along the southern component; it moved into the narrow channel separating the Great Red Spot from the South Temperate Belt, attached itself to the perimeter of the Spot and revolved around it. The direction in which the feature revolved revealed for the first time the anticlockwise "vorticity" of the Great Red Spot.

South Equatorial Belt Disturbances

Since 1919 the South Equatorial Belt has periodically faded and then dramatically returned to prominence. The revivals always begin with sudden outbursts of light and dark spots, which are localized at first, but which soon spread out in turmoil. Thirteen such disturbances have been observed, and have tended to occur at intervals of three years or else multiples of three: 1919, 1928, 1943, 1949, 1952, 1955, 1958, 1962, 1964, 1971 and 1975. The years 1943 and 1971 each saw two disturbances.

It has been found that the South Equatorial Belt Disturbances can be divided into three groups, according to their observed intensities. Moreover, although the outbreaks appear to be randomly situated, they can all be traced to three sources (corresponding to the intensity grouping) which are stationary relative to one another. These sources thus define a coordinate system of their own, whose rotational period is only 0.4 sec longer than that of System III (*see* page 24), which is thought to refer to the deep interior of the planet. It therefore seems possible that the South Equatorial Belt Disturbances have their origins in three "hot spots" located deep within the Jovian atmosphere.

23 August 1927

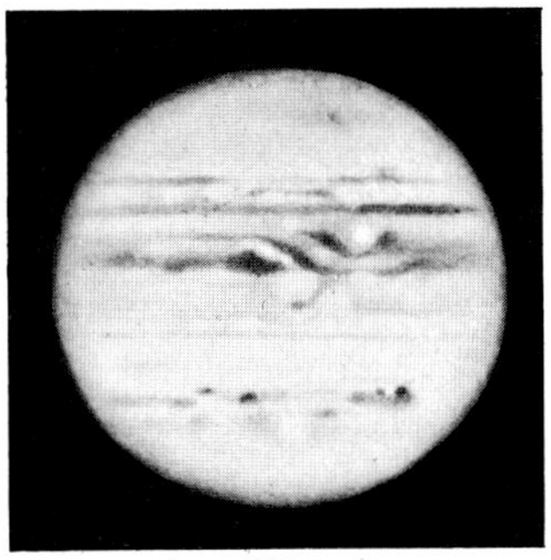

27 September 1927

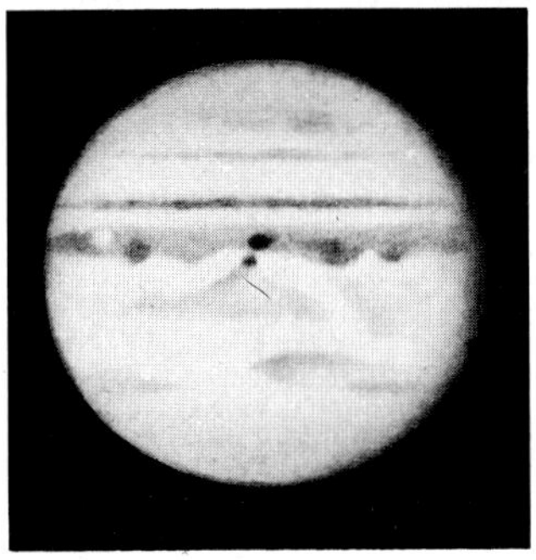

21 July 1928

27 August 1928

5 November 1928

13 February 1929

3. Historical behavior of the GRS
This graph shows how the GRS has drifted in longitude since 1931. These changes indicate that the GRS is not a fixed feature, nor is it associated with any fixed topographical feature beneath Jupiter's clouds.

4. Historical behavior of the white ovals
Like the GRS the white ovals have exhibited a tendency to drift in longitude, the average change amounting to a few degrees per year (**A**). They have also shrunk markedly since the 1930s (**B**).

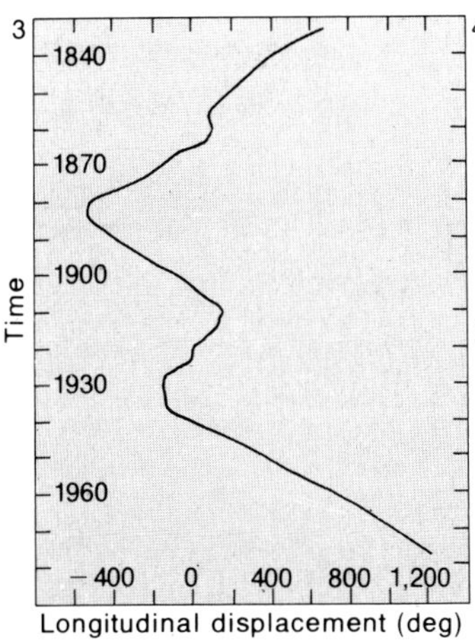

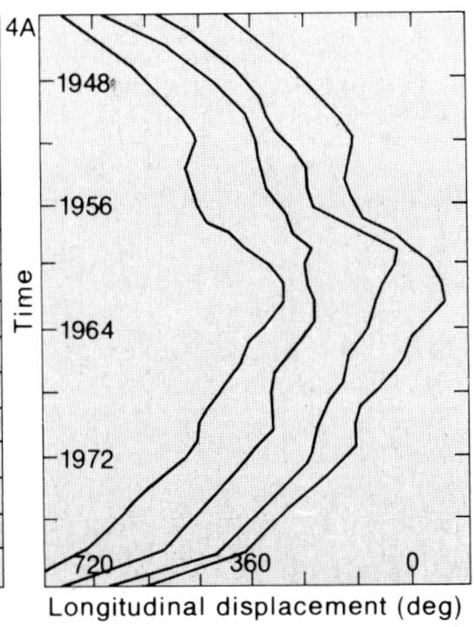

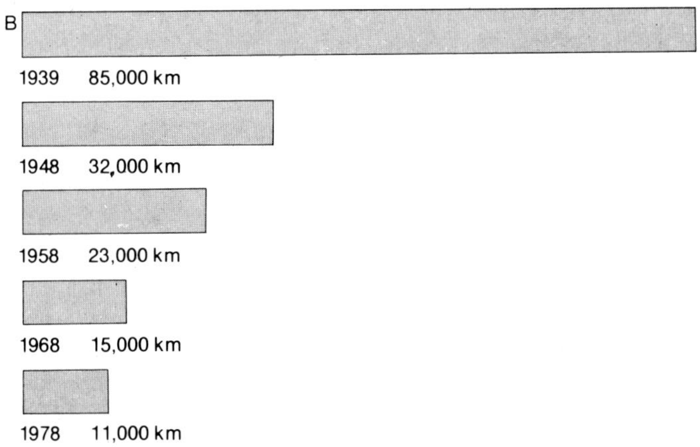

Early Theories

It was once thought that Jupiter might be a kind of miniature sun, warming its extensive satellite system, but theoretical work carried out by H. Jeffreys in 1923 and 1924 refuted the notion that Jupiter had a hot surface. Instead, Jeffreys proposed a model according to which the planet would have a rocky core, a mantle made up of ice and solid carbon dioxide, and a very deep tenuous atmosphere. This version was more or less superseded by an alternative model proposed by R. Wildt in 1934, following the positive identification of methane and ammonia in the Jovian atmosphere. According to Wildt, there was a rocky core overlaid by two shells, the first being composed of dense ice, and the second of highly compressed gases existing mostly in a condensed state.

The next widely accepted model was proposed in 1951, independently by W. Ramsey in England and W. DeMarcus in the United States. According to this model, there was a core composed of "metallic hydrogen" (that is to say, hydrogen compressed to the point at which it becomes a good conductor of electricity); around this core the model proposed a layer of liquid hydrogen, and above this the atmosphere. The present view, due mainly to the work of J. D. Anderson and W. B. Hubbard, is described on pages 18–19; it resembles the previous model in some respects, but incorporates a solid central core of iron and silicates.

Theories about the Great Red Spot

Originally the Great Red Spot was believed to be a kind of Jovian volcano: its redness was associated with heat. Before long, however, it became clear that such an explanation was untenable, and an alternative theory (sometimes called the "floating raft theory") was proposed, first by G. W. Hough in 1881, and later by B. M. Peek, who described the idea as follows:

"Consider first the well-known experiment of immersing an egg in a solution of salt and water. If the solution is more concentrated towards the bottom of the containing vessel, as it is likely to be at first, the egg, while remaining completely under water, will float at a level determined by its density. Now replace the solution by Jupiter's atmosphere, of which the density increases rapidly with the depth until it almost certainly approaches that of the liquid state, and let the egg be represented by some solid whose upper surface lies at least some tens of kilometres below the top of the cloud layer . . . Any influence tending to disturb the equilibrium of the density distribution in the atmospheric layers will bring about a change in the level at which the solid will float . . . Now one of the objections that may be raised against the hypothesis that the Red Spot may owe its appearance to the presence of such a floating solid is the enormous bulk, its length being about 40,000 kilometres, its breadth say 12,000 kilometres, and its depth presumably at least 2,000 kilometres; but if these are granted, it should be easy enough to accept the next postulate, namely that, during the last 120 years, its level has been subject to slight variations having a total range not greatly exceeding 10 kilometres. Indeed, if there were no mechanical resistance to changes of motion, the range of 10 kilometres in depth would suffice to account for all the changes in rotation period and for the whole drift in longitude of the Red Spot since 1831."

In 1963 a different theory was put forward by R. Hide. He suggested that the Great Red Spot might be the top of a "Taylor column", a relatively stagnant column of fluid which is formed over an obstacle in a rotating fluid. Such columns were first studied in the laboratory by Sir Geoffrey Taylor in 1921. Hide proposed the idea that some topographical feature on the (assumed) surface of Jupiter might be acting as an obstacle to the planet's rotating atmosphere, thus producing a "Taylor column" in the atmosphere.

This hypothesis, however, does not account for the variations in longitude of the Red Spot and, like the "floating raft" theory, it has now been rejected in view of the information sent back by the Pioneer and Voyager space-probes.

1. Theoretical model—1923
As described by H. Jeffreys, Jupiter's interior consisted of a rocky core of radius 46,000 km, surrounded by an 18,000 km thick layer of ice and solid carbon dioxide, with an atmosphere of negligible density 6,000 km deep.

2. Theoretical model—1934
This cut-away drawing illustrates the model proposed by R. Wildt. A very dense, rocky inner core of radius 30,200 km is overlaid by a 27,200 km layer of high-pressure ice, followed by a 12,600 km layer of condensed gases. This outer layer is gaseous at levels very close to the surface.

3. Theoretical model—1951
The model proposed by W. Ramsey and, independently, W. DeMarcus consists of a core of metallic hydrogen with a radius of 61,000 km, surrounded by a layer of liquid hydrogen 8,900 km deep and a shallow atmosphere.

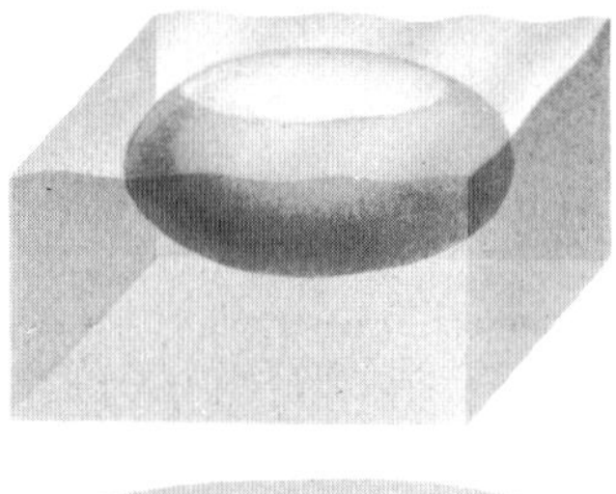

4. Early theory of the GRS
An early interpretation of the GRS, sometimes referred to as the "floating raft theory", described the Spot as a solid object floating at, or just below, the surface of Jupiter's visible atmosphere. Slight changes in its level, it was suggested, might account for its irregular drift.

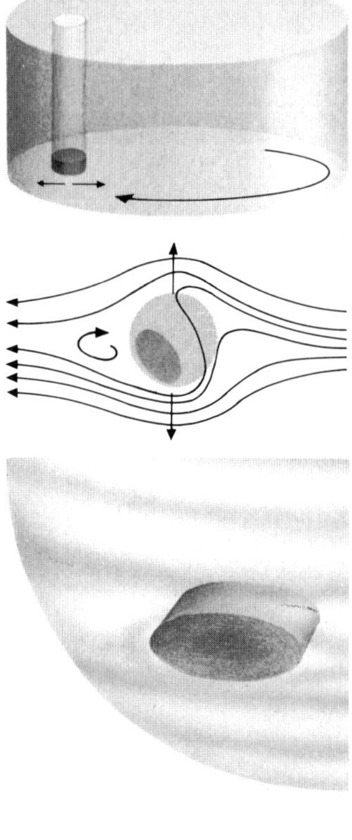

5. "Taylor column" theory
An alternative theory of the nature of the GRS was based on a study of rotating fluids. The laboratory apparatus shown in the illustration (**A**) can be used to demonstrate an effect called a "Taylor column". The tank is filled with water and rotates at high speed, while an obstacle at the bottom of the tank moves slowly towards the center; a stagnant column forms above the obstacle. When the relative speeds are correctly adjusted a Taylor column is formed vertically above the obstacle, and the flow-pattern as viewed from above (**B**) resembles the GRS in appearance. R. Hide suggested that an obstacle on Jupiter's surface might be responsible for producing the GRS by some similar mechanism (**C**). However, several objections could be made to this theory even before the results from the Pioneer and Voyager missions were available. For example, in the laboratory experiment the obstacle has to move in a direction at right angles to the flow of the liquid; but on Jupiter the latitude of the GRS has remained more or less constant over a long period.

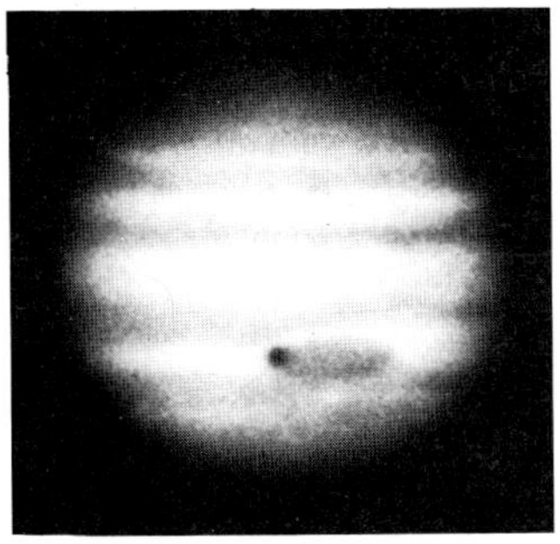

August 1891

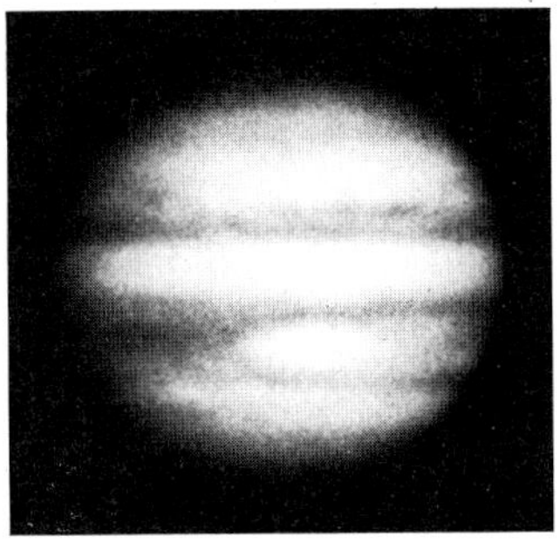

April 1908

November 1916

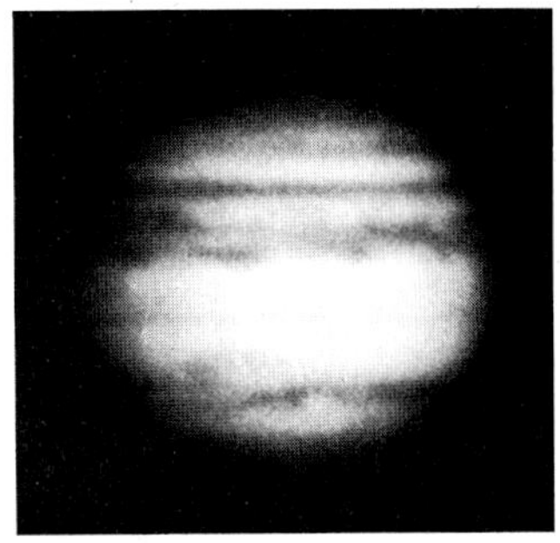

November 1927

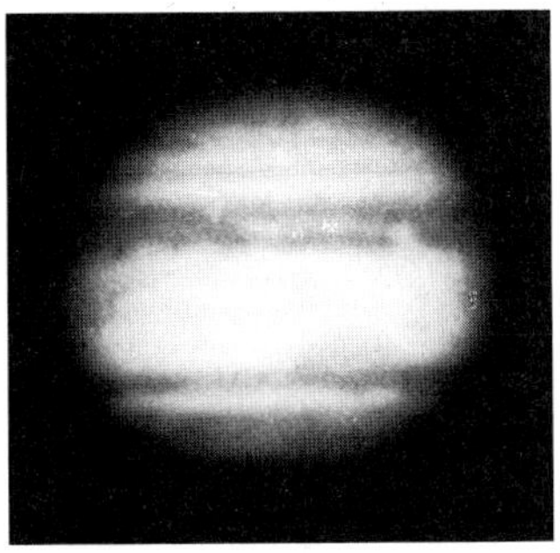

July 1937

June 1947

May 1956

November 1964

February 1966

January 1967

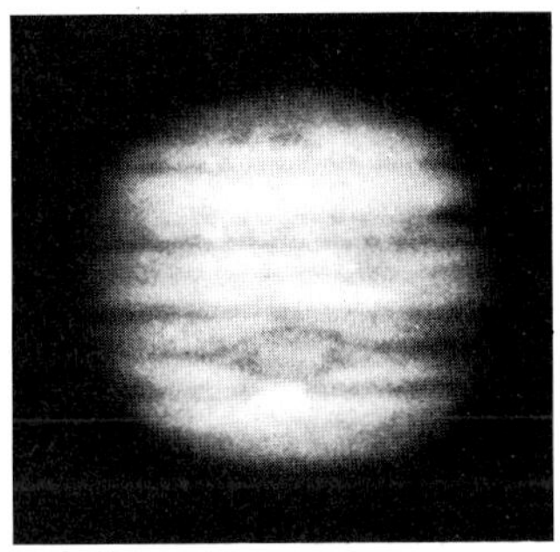

February 1968

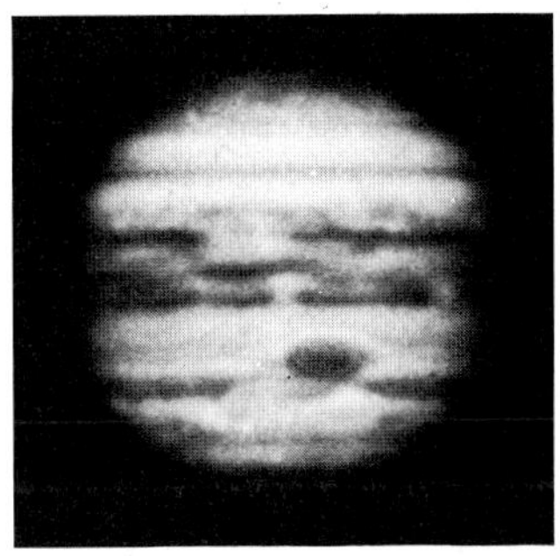

June 1972

July 1973

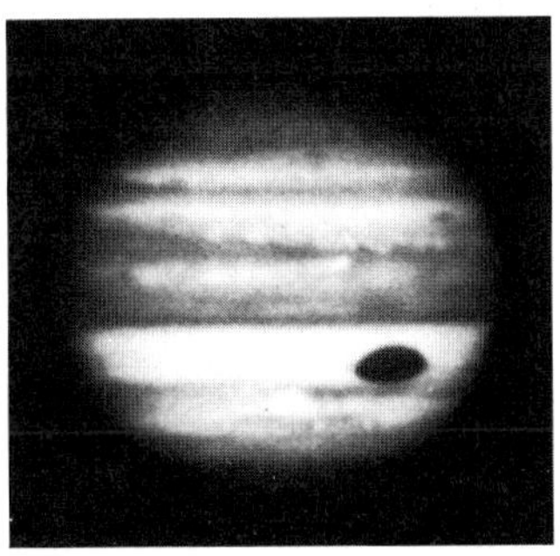

October 1974

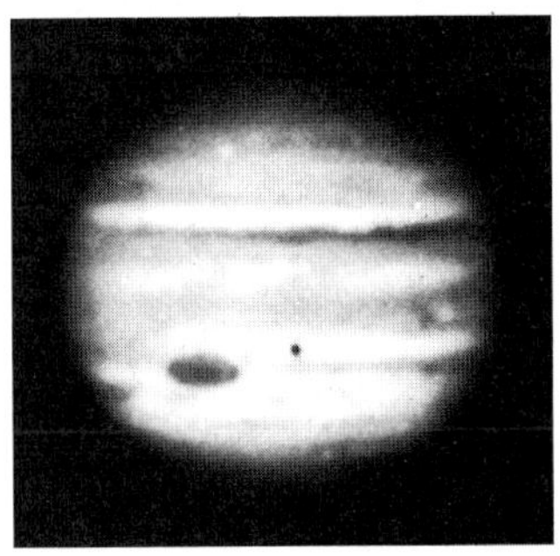

October 1975

November 1976

January 1978

December 1978

January 1979

6. Jupiter since 1891
These photographs of Jupiter reveal how the appearance of the planet has altered with time. The belts and zones fluctuate markedly in position and brightness, and the atmosphere appears to go through phases of turbulence and quiet. In the earliest photograph the GRS is nearly twice its present length, while in the following six photographs it has faded almost entirely from view.

The Voyager Spacecraft

Four vehicles have so far bypassed Jupiter. The two most recent, Voyager 1 and Voyager 2, are more advanced in design than their predecessors, Pioneer 10 and Pioneer 11, in a number of respects. In particular, their onboard computer systems are capable of directing more sophisticated experimental equipment, and the vehicles carry a more powerful source of electricity in the form of three Radioisotope Thermoelectric Generators (RTGs). While in flight, the spacecraft is stabilized along three axes, using the Sun and the star Canopus as celestial references. After the first 80 days of flight the 3.66 m parabolic reflector points constantly back to Earth.

Behind the reflector dish is a ten-sided aluminium framework containing the spacecraft's electronics. This framework surrounds a spherical tank containing hydrazine fuel used for maneuvering.

Twelve thrusters control the spacecraft's attitude, while another four are used to make corrections to the trajectory. Only trajectory changes require instructions from the ground; all other functions can be carried out by the onboard computer, in contrast to the Pioneer spacecraft which had to be flown "from the ground". There are three engineering subsystems: the Computer Command Subsystem (CCS), the Flight Data Subsystem (FDS) and the Attitude and Articulation Control Subsystem (AACS).

The entire vehicle weighs only 815 kg and carries equipment for 11 science experiments.

1

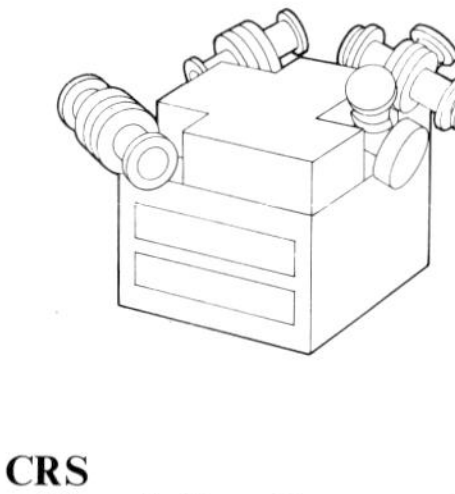

1. CRS
The Cosmic Ray Detector System is designed to measure the energy spectrum of electrons and cosmic ray nuclei. The instrument studied the composition of Jupiter's radiation belts, as well as the characteristics of energetic particles in the outer Solar System generally. The experiment uses three independent systems: a High-Energy Telescope System (HETS), a Low-Energy Telescope System (LETS) and an Electron Telescope (TET). These enabled the spacecraft to study a wide range of the particles that make up cosmic rays.

2

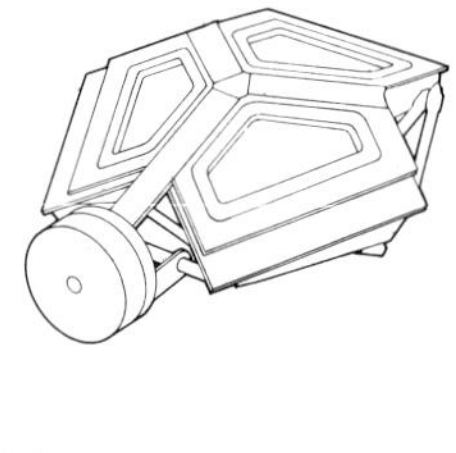

2. PLS
The Plasma experiment studied the properties of the very hot ionized gases that exist in the interplanetary regions. The instrument consists of two plasma detectors, one pointing in the direction of the Earth and the other at a right angle to the first. This equipment analyzed the properties of the solar wind and its interaction with Jupiter, as well as the magnetospheres of Jupiter and the other planets which will eventually be visited by Voyager. The PLS also provided information about the plasma environment of the satellite Io.

3

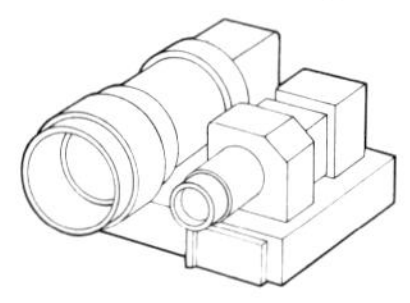

3. ISS
The Imaging Science Subsystem consists of two television-type cameras mounted on a scan platform. One of the cameras has a 200 mm wide-angle lens with an aperture of f/3, while the other uses a 1,500 mm f/8.5 lens to produce narrow-angle images. The design is a modified version of the cameras used on previous Mariner vehicles. Both cameras have a range of built-in filters as well as variable shutter speeds and scan rates. The ISS, the IRIS and the PPS instruments were able to view the same region of the planet simultaneously.

4

4. IRIS
The Infrared Radiometer Interferometer and Spectrometer measured radiation in two regions of the infrared spectrum, from 2.5 to 50 μm and from 0.3 to 2.0 μm. It provided information about the temperatures and pressures at various levels of Jupiter's atmosphere, as well as about the chemical composition of its clouds. Mounted on a scan platform, the instrument has two fields of view, one using a 0.5 m Cassegrain telescope to achieve a narrow, quarter-degree field of view, the other, pointed off the telescope sight, for a wider view.

5

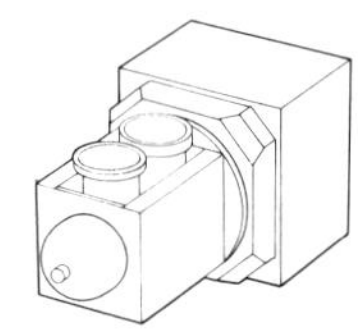

5. LECP
The Low-Energy Charged Particle experiment uses two solid-state detector systems mounted on a rotating platform. The two subsystems consist of the Low Energy Particle Telescope (LEPT) and the Low Energy Magnetospheric Particle Analyzer (LEMPA). This equipment studied Jupiter's magnetosphere (*see* page 20) and the interaction of charged particles with Jupiter's satellites. It is also designed to investigate various other interplanetary phenomena such as the solar wind and cosmic rays emanating from sources outside the Solar System.

6

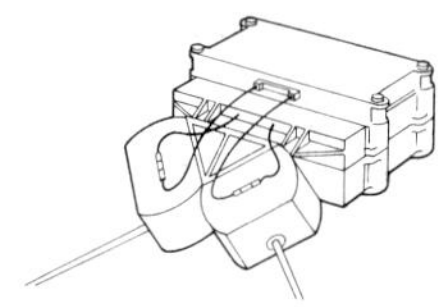

6. PWS and PRA
Two separate experiments, the Plasma Wave System and the Planetary Radio Astronomy experiment, share the use of the two long antennas which stretch out at right-angles to one another forming a "V". The PWS studied wave-particle interactions and measured electric-field components of plasma waves over a frequency range of 10 Hz to 56 kHz. This system is also designed to measure the density of thermal plasma near the planets visited by the spaceprobe. The PRA experiment detected and analyzed radio signals emitted by the planets; Jupiter was known to be a powerful source of radio waves from Earth-based experiments. The PRA receiver covers two frequency bands, the first in the range of 20.4 kHz to 1,300 kHz, and the second between 2.3 MHz and 40.5 MHz.

7

7. PPS
The Photopolarimeter System consists of a 0.2 m telescope fitted with filters and polarization analyzers and is mounted on a scan platform. It covers eight wavelengths in the region between 235 μm and 750 μm. The PPS measured gases in planetary atmospheres, examined particles present in the atmospheres, and searched the sky background for interplanetary particles. The PPS instrument is also designed to examine the surface texture and composition of the satellites of Jupiter and Saturn, and the properties of Io's sodium cloud.

8

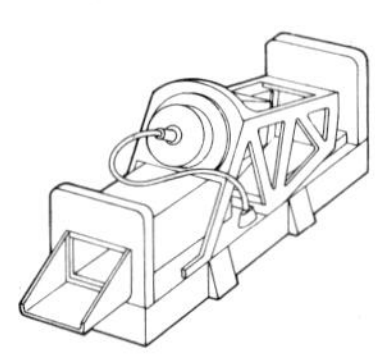

8. UVS
The Ultraviolet Spectrometer covers the wavelength range of 40 μm to 180 μm looking at planetary atmospheres and interplanetary space. Its purpose is to study the chemistry of the upper layers of the atmospheres, and to measure how much of the Sun's ultraviolet radiation they absorb during occultation; it also measures ultraviolet emissions from the planetary atmosphere. The instrument collects and channels light through a collimator, which directs a number of narrow parallel beams onto a diffraction grating.

9

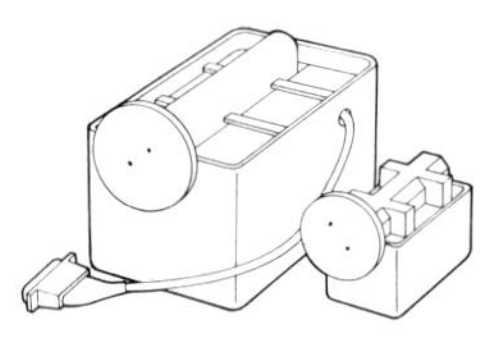

9. MAG
The Magnetic Fields Experiment consists of four magnetometers; two are low-field instruments mounted on a 10 m boom away from the field of the spacecraft, while the other two are high-field magnetometers mounted on the body of the spacecraft. Each pair consists of two identical instruments, which makes it possible to eliminate the spacecraft's field from the results. Each magnetometer measures the magnetic component along three perpendicular axes, from which the direction and strength of the field can be determined.

10. RSS
The investigations of the Radio Science System are based on the radio equipment which is also used for two-way communications between the Earth and Voyager. For example, the trajectory of the spacecraft can be measured accurately from the radio signals it transmits; analysis of the flight-path as it passes near a planet or satellite makes it possible to determine the mass, density and shape of the object in question. The radio signals are also studied at occultations for information about the occulting body's atmosphere and ionosphere.

1 High-gain antenna for communications and Radio Science Experiment (RSS) (3.7 m diameter)
2 Cosmic Ray Experiment (CRS)
3 Plasma Experiment (PLS)
4 Imaging Science (ISS)
5 Ultraviolet Spectrometer (UVS)
6 Infrared Radiometer Interferometer and Spectrometer (IRIS)
7 Photopolarimeter (PPS)
8 Low-Energy Charged Particle Experiment (LECP)
9 Hydrazine thrusters (16)
10 Micrometeorite shield (5)
11 Optical calibration target and radiator
12 Plasma Wave Experiment (PWS)
13 PWS and PRA antennas
14 Radioisotope Thermoelectric Generators (RTG)
15 High-field Magnetometer (MAG)
16 Low-field Magnetometer (MAG)
17 Electronics compartments
18 Fuel tank

Trajectories and Communications

The first spacecraft to reach Jupiter was Pioneer 10, launched on 2 March 1972, which made its closest approach to the planet on 3 December 1973 at a distance of approximately 132,000 km. Pioneer 11 was launched on 5 April 1973 and bypassed Jupiter on 2 December 1974, almost exactly a year after its predecessor.

Pioneers 10 and 11 were virtually identical, but Pioneer 11 made a closer approach to the planet (42,800 km) and passed more rapidly over the equatorial zone, following a trajectory which took it over Jupiter's south pole. It was then decided to make a course adjustment which would take Pioneer 11 back across the Solar System towards a rendezvous with Saturn on 1 September 1979. This extended program was successfully carried through, providing the first close-range information from Saturn. Both Pioneers are now travelling out from the Solar System, never to return.

In 1977 a rare alignment of the outer planets (once in 176 years) offered the possibility of sending space-probes on a "Grand Tour", taking them in turn past Jupiter, Saturn, Uranus and possibly Neptune as well. Because of the immense distances involved and the limited supply of fuel that can be carried on board, a special trajectory known as "gravity-assist" is used to boost the vehicle on its journey. The principle is to use the gravitational field of each planet to accelerate the spacecraft on to the next.

The two Voyagers were launched in the late summer of 1977. Voyager 2 was in fact launched first (20 August) and made its pass of Jupiter on 9 July 1979 at 714,000 km. It is scheduled to reach Saturn in August 1981, Uranus in January 1986 and possibly Neptune in 1989. Voyager 1, launched on 5 September 1977, travelled by a more economical route and reached Jupiter first, passing it at 350,000 km on 5 March 1979. Voyager 1 went on to reach Saturn in November 1980.

New probes have already passed the design stage. Of particular importance is Project Galileo, due to be launched in 1985, which is intended to send a probe into the Jovian atmosphere.

1

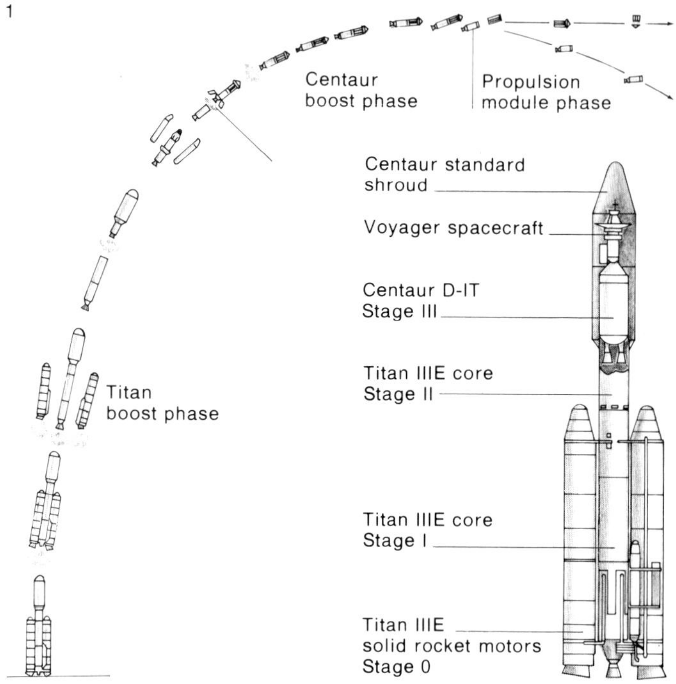

1. Titan/Centaur
The Voyager spacecraft were launched from Cape Canaveral by a vehicle consisting of the Titan III-E and a Centaur D-IT upper stage, with a protective assembly called the Centaur Standard Shroud (CSS). The thrust at lift-off was provided by two Solid Rocket Motors (SRMs). The launch involved a total of six engine burns, the last provided by a propulsion module suspended below the mission module.

2. Voyager trajectories

2

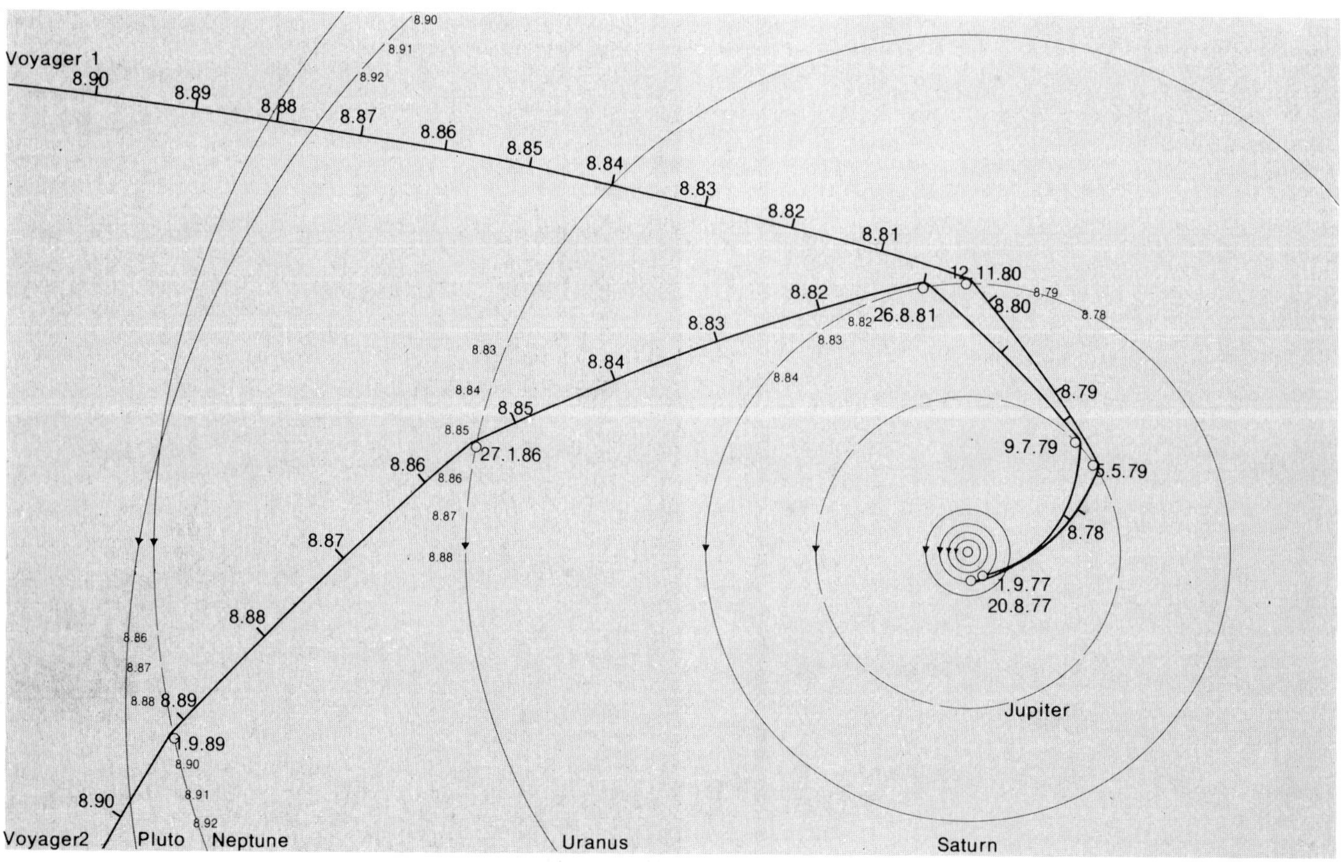

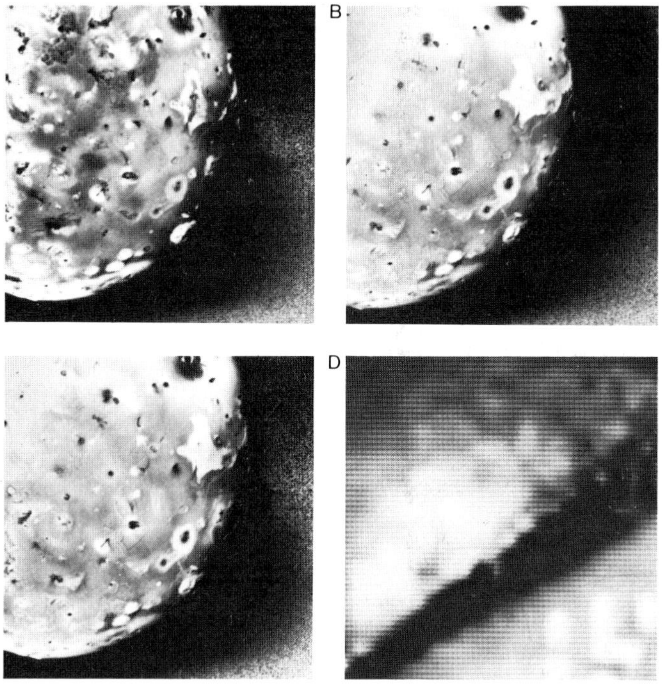

3. Image transmission
Full-color images are built up from monochromatic images taken through different filters, such as violet (**A**), orange (**B**) and green (**C**). Color balance can be improved later as part of the computer processing to produce the final image. Colors may also be enhanced to bring out hidden details. A magnified view (**D**) reveals the individual pixels.

The problems involved in tracking, controlling and communicating with space-probes as far away as Jupiter are considerable. The power of the onboard transmitters is severely limited by the restriction on weight; those on the Voyager spacecraft operated on a maximum of less than 30 W, and the power of the received signal over 1 m^2 of the Earth's surface was in the range of only 10^{-18} W. Moreover, because of the immense distances, there is a delay of about 40 min between the transmission and reception of a signal.

Also, a certain amount of interference and distortion is inevitable. Some types of information, such as command signals, demand a high degree of accuracy; others, such as video data, are relatively tolerant, since errors can be removed by computer-processing on the ground. Sophisticated encoding techniques are used to protect the most error-sensitive data, while at the same time more robust data can be transmitted more economically. The communications system operated on two different frequencies, one near 0.13 m in the so-called "S-band" and another near 0.4 m—the "X-band". The X-band was used exclusively for "downlink" transmissions, while the S-band carried both "uplink" and "downlink" data. A network of three space stations, in Canberra, Madrid and Goldstone, tracked the Voyager spacecraft; these stations are sited at such longitudes that the spacecraft is within range of one at all times.

The transmission of images is of particular interest, and in this respect the Voyagers represent a significant advance over the Pioneers. In flight the Pioneer vehicles were constantly spinning about one axis, and their cameras built up an image gradually, scanning a narrow strip during each rotation. The Voyagers, on the other hand, employ shuttered television-type cameras. Each image is analyzed into 640,000 picture elements or "pixels"; the brightness of each pixel is measured, and its value expressed as an eight-digit binary number, thus converting the image into a long string of binary code that can either be transmitted immediately or stored in the memory, whose storage capacity is equivalent to 100 pictures.

4

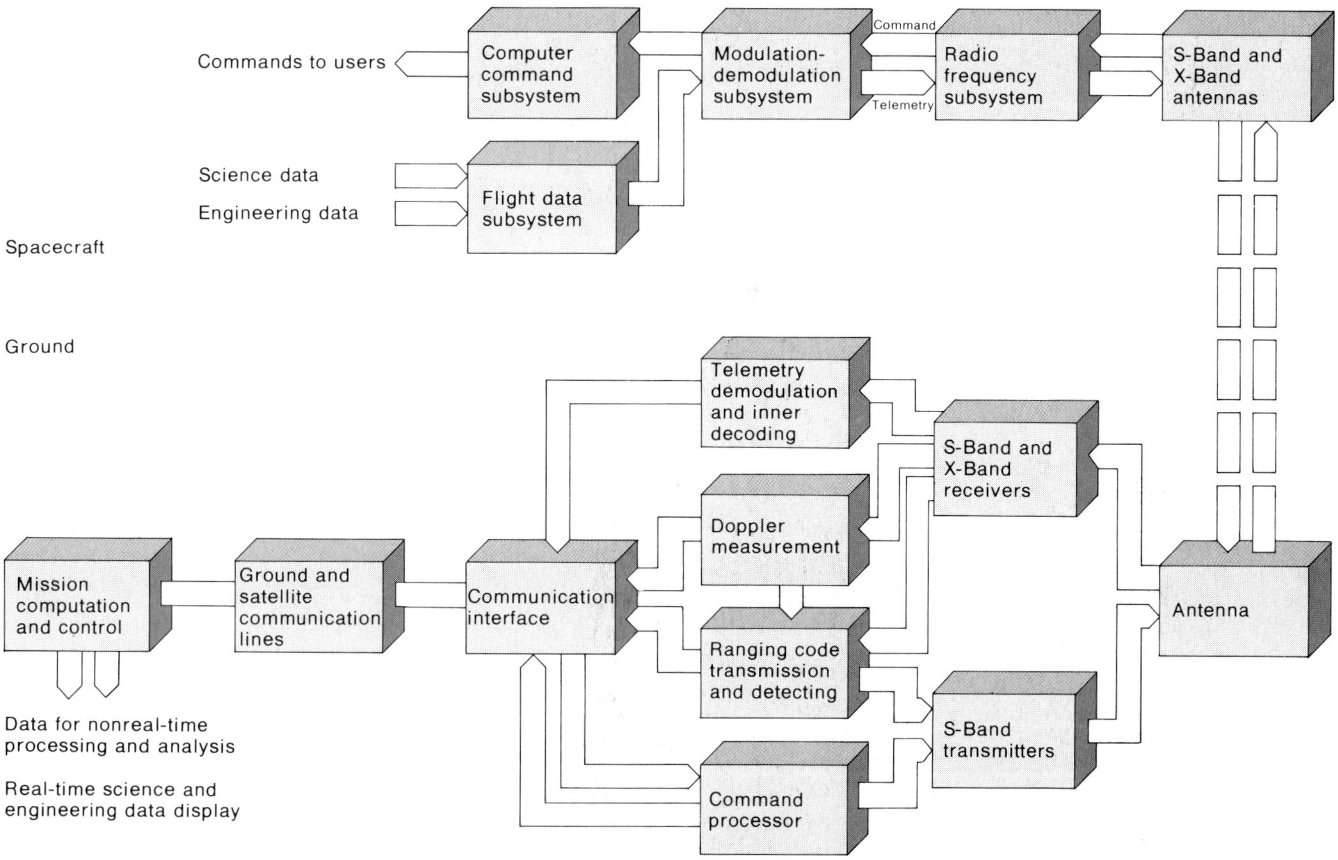

4. Telecommunications system
This block diagram illustrates the stages of data transmission.

Structure of Jupiter Interior

Planetary magnetism

The first evidence that Jupiter had a magnetic field came in the mid-1950s through the unexpected detection of radio waves from the planet by ground-based radio telescopes. The intense, fluctuating radio emission could only be interpreted in terms of energetic particles interacting with a magnetic field around the planet. The field and belts have now been studied directly by the Pioneer and Voyager spacecraft.

The only other planets currently known to possess magnetic fields are Mercury, the Earth and Saturn. The discovery of a very weak field around Mercury was a great surprise, and its origin remains unknown. The Saturnian field was found from measurements made by the Pioneer spacecraft in September 1979.

The magnetic field that has been subjected to the most intense study is the Earth's, and it is thought to originate from within the planet's core. The mechanism generating the field may be compared with a dynamo, which is a machine for converting mechanical energy into electrical or magnetic energy. In some ordinary dynamos there is a built-in permanent magnet which supplies the magnetic field, but it is possible to construct them in such a way that they generate their own fields: these are called "self-exciting" dynamos and illustrate the principle that probably underlies all the planetary magnetic fields that have been observed (*see* diagram 1). The two essential features are the presence of a good electrical conductor and a source of mechanical energy to drive the dynamo.

In the case of a planet, the electrical conductor is thought to be some electrically conducting fluid at or near to the planet's core; but the force driving the system is not known and may involve a combination of motions made up of several factors. The Earth, for example, is believed to have a metallic core composed principally of iron, which may be solid in the center but which is probably dense liquid in the outer regions. Various suggestions have been made to account for movements within the fluid zone sufficient to create the observed magnetic field, including convection brought about by the flow of heat outward from the core and the motions of the Earth's rotation and orbit.

Interior

It is clear that the origin of the Jovian magnetic field is closely related to the structure of the planet's interior; indeed the existence

2

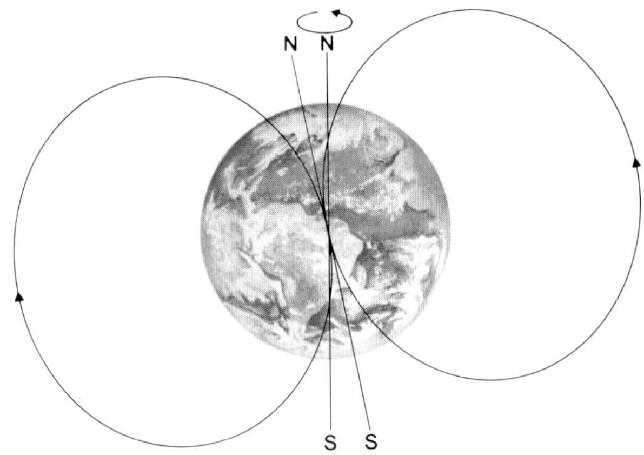

1. The principle of the dynamo
A simple model of a disc dynamo consists of a metal disc rotating in a magnetic field between two permanent magnets (**A**). The field produces a force on the free electrons in the disc, pushing them towards the center. As a result there is a difference in electrical potential between the edge and the center of the disc, which would produce a current if the circuit is closed. In a self-exciting dynamo (**B**) this current is used to drive an electromagnetic coil, which replaces the original permanent magnets. The resulting system generates a magnetic field as long as the disc is kept spinning. The model thus demonstrates how mechanical energy may be converted into magnetic energy; some analogous process is thought to be responsible for creating planetary magnetic fields.

2. The Earth's magnetic field
At present the Earth's magnetic field is inclined to the axis of rotation by approximately 11°, the same angle as on Jupiter. On Earth the magnetic north is, at present, in the same direction as the geographic north. However, it is known that in the past the Earth's magnetic polarity has reversed itself, and it is conceivable that similar reversals occur on Jupiter.

3. The Earth's interior
Most of the information on the Earth's interior is derived from the study of shock waves such as those generated by earthquakes. These have revealed a crust extending to a depth of between 10 and 40 km, which covers a dense shell called the mantle. The mantle is about 3,000 km thick, and below it is a 2,000 km thick outer core surrounding a solid inner core.

1A

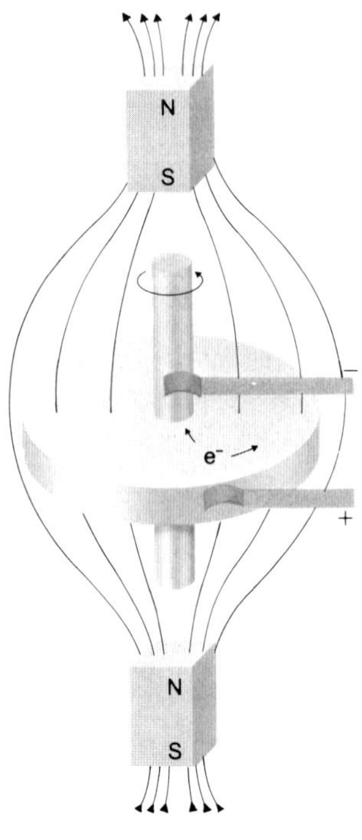

B

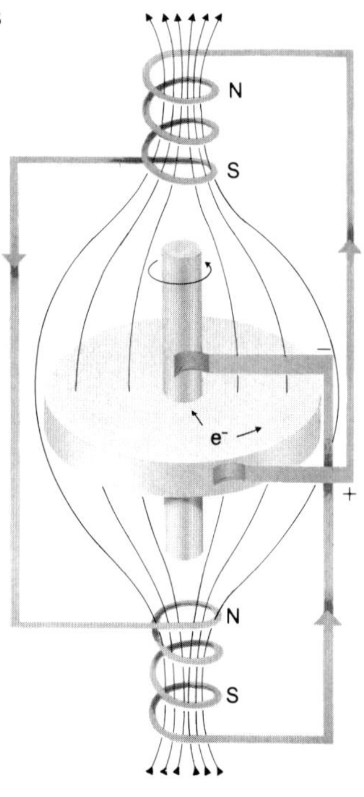

3

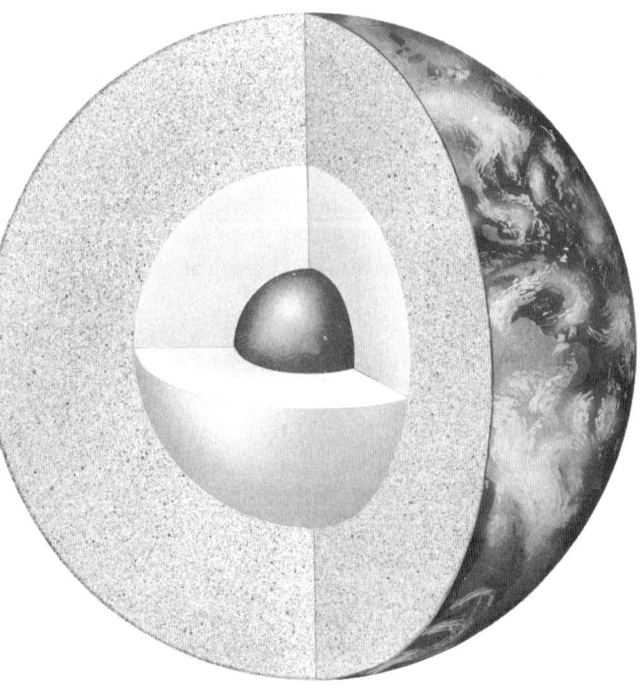

of a magnetic field around Jupiter, and similarly around Saturn, may be interpreted as evidence that these planets contain zones of liquid, although there is no direct evidence in either case. Jupiter's electrically conducting core may be as large as 0.7 or 0.8 Jupiter radii (written as R_J).

Jupiter, like Saturn, is known to be composed primarily of the light elements hydrogen and helium. This fact is deduced from a knowledge of Jupiter's density, which is approximately 1.3 times that of water; it is thus apparent that Jupiter must be radically different in composition from the so-called "terrestrial" planets, whose mean densities are in the region of five times the density of water. The giant planets have gravitational fields powerful enough

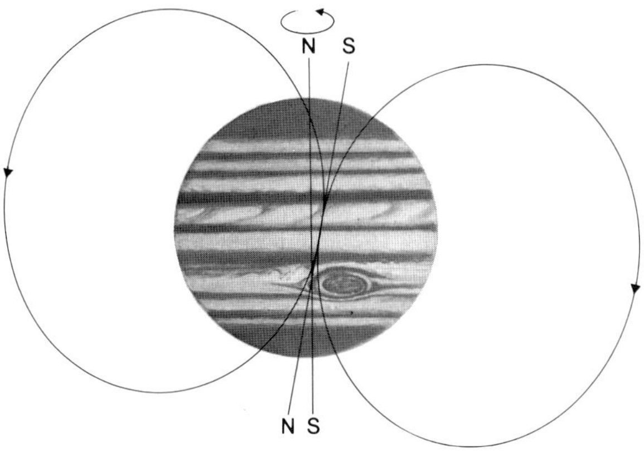

4. Jupiter's magnetic field
The magnetic field of Jupiter is inclined to the planet's axis of rotation by an angle of about 11°. Although the field is more complex than that of the Earth, it may be thought of as behaving as though a bar-magnet were embedded at the planet's core. The alignment of north and south magnetic poles is such that a terrestrial compass needle would point south.

5

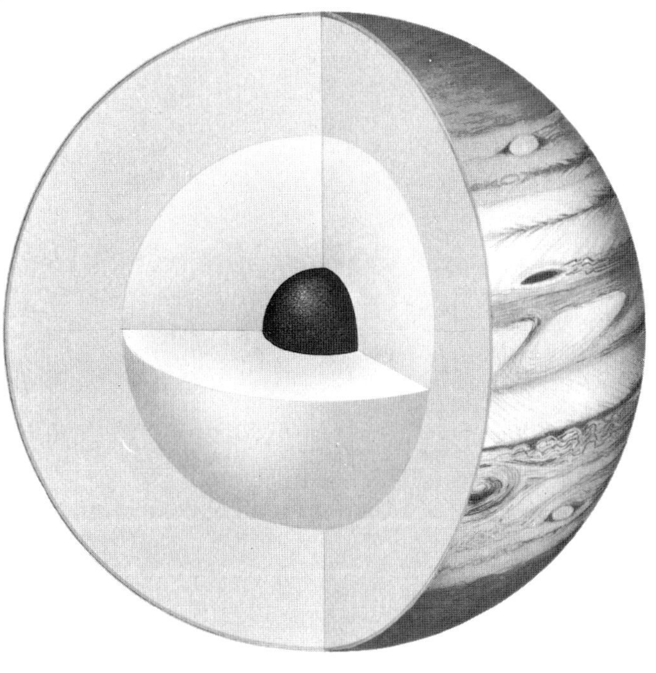

5. Jupiter's interior
Jupiter is composed almost entirely of gases and liquids, although at present it is believed that there is a small core of rocky materials. The innermost part of the planet around this core is thought to consist of liquid metallic hydrogen, surrounded by a shell of liquid molecular hydrogen. The upper region consists of a deep atmospheric layer of hydrogen and helium.

to retain these light elements, and models of their interiors depend heavily on the assumed proportions of hydrogen to helium.

Its interior is now believed to have the following structure: at a depth of about 1,000km from the visible surface there is a transition zone from gaseous to liquid hydrogen. The temperature at this level is about 2,000 K and the pressure about 5,600 Earth atmospheres (that is to say, 5,600 times the pressure of the Earth's atmosphere at sea-level). Further down, at a level of 3,000 km, the temperature rises to 5,500 K and the pressure to 90,000 atmospheres, and as a result the hydrogen is highly compressed. Twenty-five thousand km below the cloud tops, a distance of more than a third of the radius of the planet, the temperature is over 11,000 K and the pressure is about 3 million atmospheres. In these conditions hydrogen undergoes a dramatic change, rather like the change from a gas to a liquid, altering from its "liquid molecular" form to a state in which it is a good conductor of electricity; in this state it is known as "liquid metallic hydrogen".

From this level onward, the temperature and pressure continue to rise steadily, reaching 30,000 K and 100 million atmospheres at the center. Finally, at the center of the planet, there is thought to be a small rocky core of 10 to 20 Earth masses, composed of iron and silicate materials. The likelihood of the existence of such a core is deduced from what is known about the balance of elements in Jupiter's interior.

Magnetic field

The presence of liquid metallic hydrogen in the inner regions of the planet would be of critical importance in explaining the existence of a strong magnetic field because of the material's properties as an electrical conductor. Jupiter's rapid 10 hr rotation might be wholly or partly responsible for setting up a flow of liquid to generate electric ring-currents on the same fundamental principle as the dynamo described earlier. The electric currents create magnetic fields, which in turn influence the strength of the electric currents. In fact, the magnetic fields thus produced will act in such a way as to oppose the force that creates them, limiting the strength of the field and preventing it from increasing indefinitely.

Jupiter's magnetic field has a strength of 4.2 Gauss (*see* page 92) measured at the cloud tops. It is therefore more than ten times as strong as the Earth's field of 0.35 Gauss at the surface. The Jovian field is inclined to the axis of rotation by an angle of 10.8, fractionally less than the degree of tilt found on the Earth; and the magnetic axis is displaced from the center of the planet by about 0.1 R_J, mainly along the equator.

Beyond a distance of about $3R_J$ (measured from the center of the planet) the major component of Jupiter's field is "dipolar", like that of the Earth: in other words, the field behaves as if there were an immensely powerful bar-magnet embedded inside the planet inclined to the axis of rotation and displaced from the center. At the present time the "polarity" of the field (the alignment of the north and south poles) is opposite to that of the Earth, so that a terrestrial compass taken to Jupiter would point south. However, studies of the residual magnetism in certain types of rock have shown that the polarity of the Earth's field undergoes periodic reversals, at intervals of approximately 2×10^5 years (on average) for a complete reversal; it is possible that the same phenomenon occurs on Jupiter, though the time-scale of any Jovian field reversal is unknown.

Closer to Jupiter, however, there are some important differences in the structure of its magnetic field as compared with that of the Earth. Although the principal component remains dipolar, the simple model of a bar-magnet is no longer very accurate and must be replaced by a complex system of quadrupole and octopole moments, which act rather like harmonic frequencies superimposed onto a pure musical note. The cause of these high-order magnetic moments may be the complicated circulation patterns that take place in the metallic hydrogen interior.

Structure of Jupiter Magnetosphere

Magnetic environment
Jupiter's magnetic field continues to influence the behavior of charged particles at a distance far above the planet's atmosphere: the region in which its influence is dominant is called the "magnetosphere". Despite its name, the magnetosphere is not in fact spherical, but has a long "magnetotail" streaming away from the direction of the Sun, stretching out almost 750 million km away, beyond the orbit of Saturn. The Earth's magnetosphere resembles that of Jupiter in structure, but is very much smaller. Jupiter's magnetosphere is immense; if it were visible from the Earth, the spherical part would occupy as much sky as the Sun. Its size is, however, variable, extending to a distance of between 50 and 100 R_J in the direction of the Sun. By contrast, it is extremely rare for the Earth's magnetosphere to vary much in size.

Changes in the size of Jupiter's magnetosphere are brought about by its interaction with a continuous stream of particles, called the "solar wind", emanating from the Sun. The particles that make up the solar wind are mainly electrons and protons, and the strength of the wind depends on the level of activity on the Sun, where magnetic storms take place over an 11-year cycle. When the solar wind particles collide with the magnetosphere, they are abruptly slowed down from speeds of approximately 1,500,000 km hr^{-1} to a mere 400,000 km hr^{-1}. This rapid deceleration causes a huge rise in the effective temperature, which is increased by as much as a factor of ten. (The term "effective temperature" is used because it depends on the idea that the kinetic energy of particles due to their random movement can be interpreted as heat energy: but in the highly rarefied regions that exist here, the individual particles are spaced so widely apart that although they may be highly energetic individually, the total kinetic energy in a given volume of space amounts to relatively little heat. Thus spacecraft can safely pass through regions in which the effective temperature is higher than in any part of the Sun.)

The collision between the solar wind and Jupiter's magnetic field forms a shock wave, called the "bow shock", within which there is a turbulent region called the "magnetopause". Just inside the magnetopause the Voyagers measured temperatures of 300–400 million K, the highest found in the Solar System. Finally, the envelope enclosing all the magnetically active regions is known as the "magnetosheath".

Equilibrium is established between the external pressure of the solar wind and the internal pressure of the magnetosphere, which is sensitive to relatively minor changes in the solar wind intensity: the stronger the solar wind, the more compressed the magnetosphere in the direction of the Sun. One of the main reasons why the Earth's magnetosphere is so much smaller than that of Jupiter is that it is much closer to the Sun, and is therefore subjected to a stronger solar wind (in fact, about 25 times stronger). The difference in size

1. Jupiter's magnetosphere
A vast region of space surrounding Jupiter is dominated by the planet's magnetic field. This region, the magnetosphere, has a complex structure. The bow shock is situated where the particles of the solar wind collide with the magnetosphere. The turbulent region immediately inside the bow shock is called the magnetopause, and the whole magnetically active region is enveloped by the magnetosheath. A current sheet of trapped plasma follows close to the magnetic equator; radiation belts dominate the innermost regions.

1

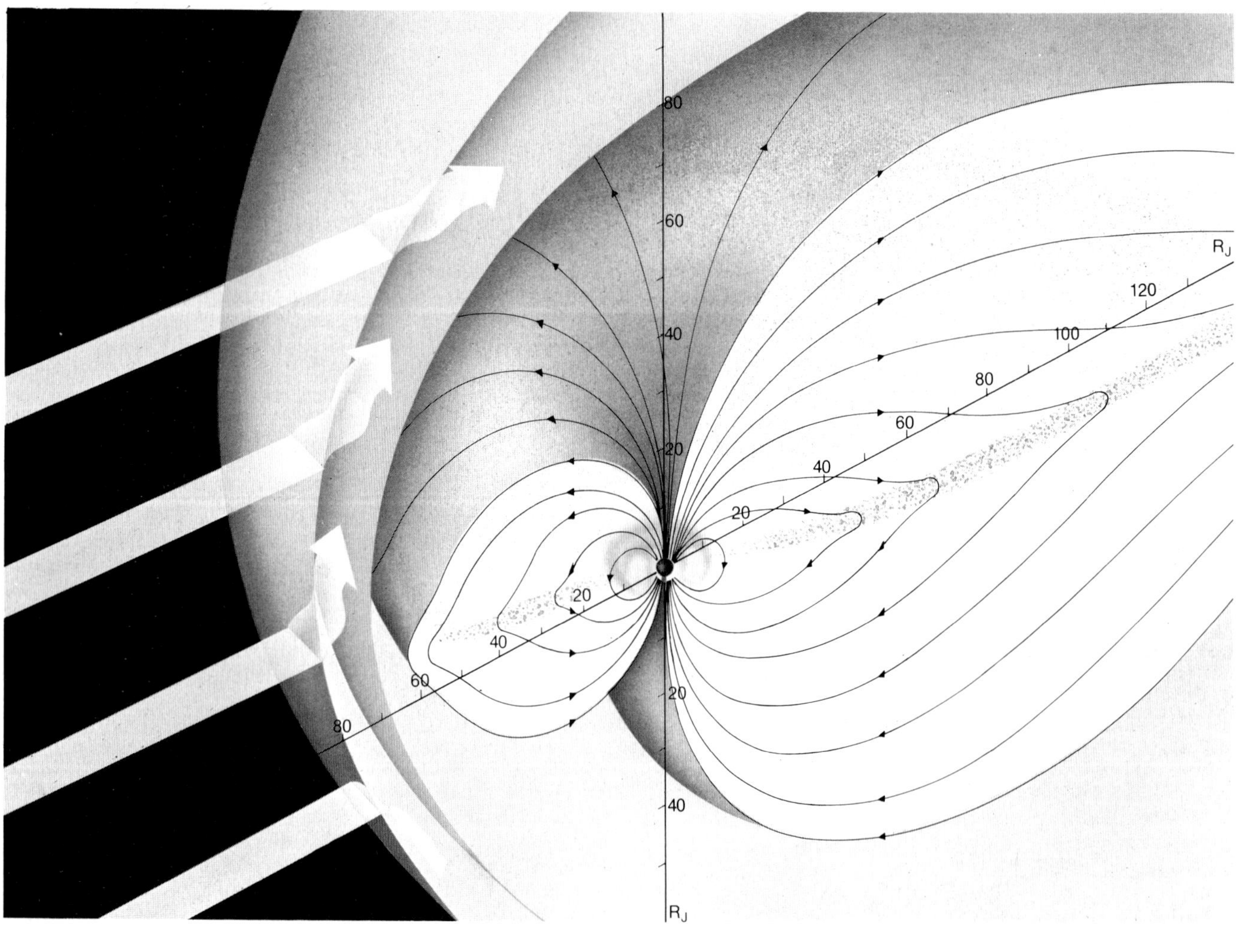

between the two magnetospheres is further explained by the fact that the Earth's field is intrinsically more than ten times weaker than Jupiter's.

In the outer regions of Jupiter's magnetosphere, beyond about 20 R_J, the magnetic field is more or less confined to a single plane which cuts the equator of the planet. Within this disc-like region flow electric currents, carried by low-energy "plasma" trapped in the field. (Plasma is matter which has acquired enough heat not only to break the bonding forces between its molecules, but also to free the electrons from their atoms: it has thus reached a state beyond the condition of existing as a gas and, because it is made up of charged rather than uncharged particles, is able to conduct electricity. The charged atoms are known as "ions".)

The electric currents form a "current sheet" which is warped in opposite directions on either side of the planet, so that it lies above the equatorial plane on one side and below it on the other. It rotates with the planet more or less like a rigid body, producing an up-and-down motion rather like a warped record, which the spacecraft periodically crossed.

The magnetosphere's mixture of trapped particles also changes rapidly. Voyager 2 detected only about a tenth as many high-speed carbon and sulphur ions as Voyager 1, but at the same time it found a higher carbon to oxygen ratio. Ten-hour variations in electron intensity have also been detected inside the magnetosphere, and similar variations have been measured as far as 150 million km from Jupiter: both inside and outside the region the variations are locked together in phase, so that it appears as though Jupiter emits these particles in a rotating beam.

Radiation belts

High-energy particles are trapped in Jupiter's magnetic field and form belts of intense radiation aligned with the magnetic axis. These belts resemble the Van Allen belts of the Earth, but are 10,000 times more intense at their highest levels. The peak intensities are found within 20 R_J, and Pioneer 10, the first spacecraft to fly through this region, received an integrated dose of 200,000 rads from electrons and 56,000 rads from protons. (For a man, a whole body dose of 500 rads would be fatal.) The radiation seriously jeopardized the spacecraft mission, and the second Pioneer vehicle followed a trajectory which took it over the south pole of the planet, passing rapidly through the most dangerous regions. Nevertheless it was found that there was a greater "flux" or flow of energetic particles at the higher Jovian latitudes than would have been expected from the measurements made by Pioneer 10 alone. Furthermore, it would appear that the flux of energetic particles reaches a maximum to either side of the dipole magnetic equator.

These radiation belts are a serious potential hazard to spacecraft. Intense bombardment by high-energy protons and electrons can saturate the sensitive experimental equipment on board, upset the functioning of the spacecraft's computer systems, and interfere with its communications with the Earth. Although the Pioneer vehicles survived, the Voyagers were carefully constructed to withstand twice the expected dose of radiation. Only minor effects were noted when the first Voyager plunged to within 350,000 km of the planet on 5 March, but Voyager 2, passing Jupiter at the much greater distance of 650,000 km, found the radiation environment three times stronger.

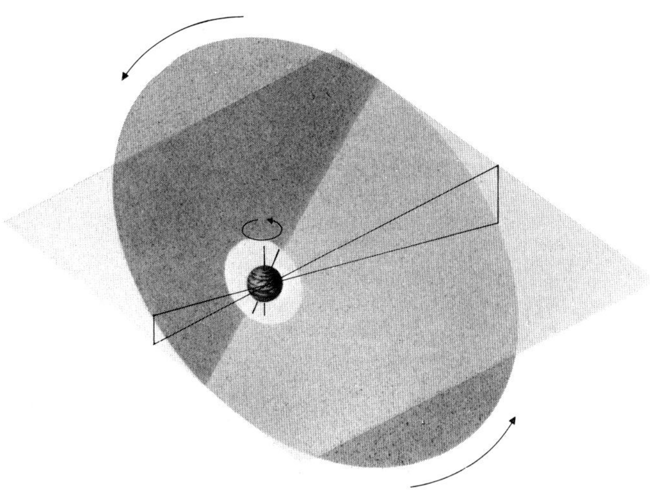

3

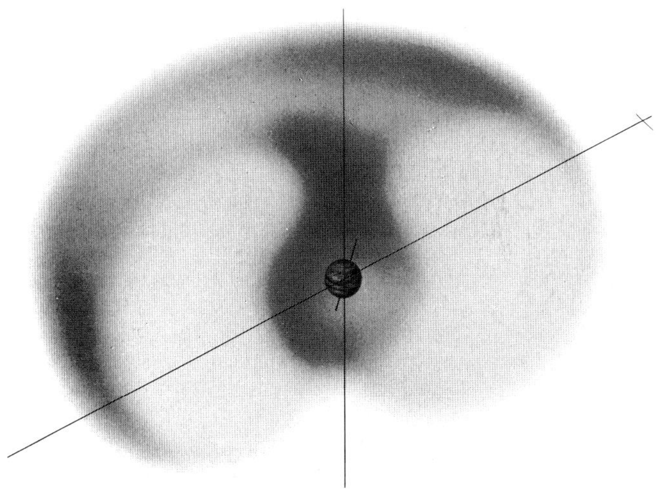

4

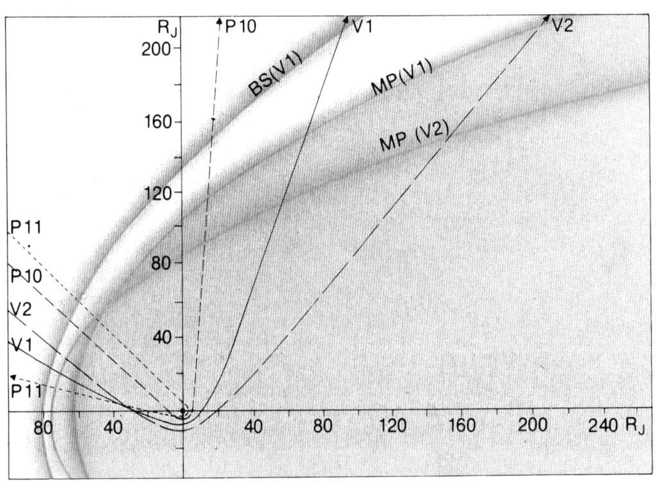

2. Current sheet
A disc-like sheet of low-energy plasma is trapped in Jupiter's magnetic field, lying roughly in the plane of the magnetic equator. The charged plasma particles carry electric currents, and the sheet rotates with the planet. Because the magnetic axis is tilted with respect to the axis of rotation, the current sheet appears to wobble up and down as it rotates. Its behavior can be compared to a warped record on a rotating turntable.

3. Radiation belts
Within about 20 R_J there exist belts of intense radiation similar to the Van Allen belts of the Earth. These belts are made up of high-energy particles and, like the current sheet, they are aligned with the magnetic rather than the rotational axis.

4. Spacecraft trajectories
The size of the magnetosphere is known to vary considerably. Each of the vehicles encountered the bow shock at different distances from Jupiter. Voyager 2, for example, crossed the bow shock at least 11 times within three days during its approach to Jupiter. Some of the points at which the spacecraft crossed the bow shock are illustrated in the diagram: curves representing Jupiter's magnetopause and bow shock are shown in three different positions, as calculated from Voyager 1 and Voyager 2 data. These curves are based on average values.

Interaction of Jupiter and Io

Cosmic rays

Sometimes Jupiter's magnetic field releases bursts of its trapped particles in the form of cosmic rays, the most energetic particles found in nature. The speed of such particles approaches that of light. They are generally atomic nuclei (protons and neutrons) stripped of their electrons, but other types of particles are also found, and the range of types and energy levels indicates that a variety of astronomical sources and processes contribute to the cosmic rays reaching the Earth. Originally it was thought that most cosmic rays came from outside the Solar System, but today it is believed that a substantial proportion are spun off Jupiter. Cosmic rays have been detected as far away as Mercury, a distance of more than 700 million km from Jupiter.

Cosmic rays could pose a hazard to future space travellers. However, although they can cause mutations in living organisms by altering or destroying genes, it is extremely unlikely that life on Earth can be endangered by them since they cannot easily reach the Earth's surface. Nevertheless, it has recently been shown that some computer systems have been affected by extremely random changes within their micro-circuits; the odds that such effects will take place are very small indeed, but it is not impossible for daily life to be indirectly influenced by the presence of cosmic rays.

Plasma torus and flux tube

Jupiter's eight innermost satellites (*see* pages 50–83) and its system of rings (*see* pages 56–57) all reside inside the hostile environment of Jupiter's magnetosphere, with the furthest of these satellites, Callisto, orbiting near the edge. In this respect, Jupiter differs markedly from the Earth, whose Moon resides outside the main body of the magnetosphere, only passing through the magnetotail as it orbits the planet. Jupiter's satellites are constantly bombarded by the high-energy electrons, protons and other particles that lie in their paths, and the interaction between the satellites and these particles results in a gradual erosion of the surfaces of the satellites. The interaction is particularly vigorous on Io (one of the four major satellites—*see* pages 58–65) and the resulting "sputtering" action is thought to be responsible for the creation of an uncharged or "neutral" atmosphere of sodium, potassium and magnesium that has been observed forming a cloud which stretches out for some distance along Io's orbit around Jupiter. Its vertical extent is about $2R_J$ (*see* diagram 3).

Theoretical ideas suggest that Jupiter and Io are connected by a flux tube of electrons and ions carrying a massive current of 5 million amps at a potential difference of 400,000 volts. The power contained in this torrent of electricity (2×10^{12} watts) is some 70 times the combined generating capacity of all the nations on Earth. This electrical energy may also play a further important role in locally heating the surface of Io, and therefore in assisting the volcanic activity of the satellite (*see* pages 62–63). The material ejected from Io's volcanoes is sent up into the atmosphere as vast

1. Io's flux tube and torus
A doughnut-shaped torus surrounds Jupiter along the orbit of Io; it consists of charged plasma particles, and since these are influenced by Jupiter's magnetic field the position of the torus is slightly inclined to Io's orbit. Io is linked to Jupiter by a flux tube carrying an immense current of 5 million amps flowing between Io and Jupiter.

1

plumes of dust and sulphur dioxide which become ionized, forming a plasma ring or torus around Jupiter in the orbit of Io.

At the time of the Voyager 1 encounter, observations were made of ionized oxygen and singly or doubly ionized sulphur (atoms of sulphur which have lost either one or two of their electrons) with a plasma temperature of 10^5 K. These observations indicate a considerable change in the Jovian environment since the Pioneer missions four and a half years earlier. A further change was found by the time of the second Voyager encounter: ultraviolet (UV) emissions from the Io plasma torus had doubled in brightness, and the temperature had decreased to 6×10^4 K.

Since the material in the torus comes from the Ionian volcanoes, changes in the properties of the torus reflect corresponding changes in the volcanic activity of the satellite. Approximately 10^{10} ions of sulphur and oxygen $cm^{-2} s^{-1}$ are required to be pumped into the torus to maintain it; the lifetime of the constituent material is estimated to be 10^6 sec. The torus material is probably transported by diffusion both inward to Jupiter and outward into space.

Aurorae

Before the Voyager mission it was thought that some solar wind particles would leak through the tail of the magnetosphere and into the polar regions of Jupiter, creating the type of auroral displays familiar on Earth, such as the Aurora Borealis. Auroral displays were indeed found, but they are in fact triggered not by the precipitating solar wind particles, as on Earth, but by the interaction between electrons, streaming in from Io's torus, and the complex Jovian ionosphere (*see* pages 26–27). Auroral activity had been virtually absent during the Pioneer flybys in 1973 and 1974.

The brightness of aurorae is generally specified in units called "Rayleighs", 1 Rayleigh being equal to 10^6 photons $cm^{-2} s^{-1}$. On Earth an aurora with a brightness of 1,000 Rayleighs is just visible, while displays of brightness of 40,000 Rayleighs are clearly visible and common enough to appear nightly in some polar latitudes. On Jupiter, on the other hand, the aurorae are even more energetic, with a brightness at UV wavelengths of 60,000 Rayleighs. (The shift towards the ultraviolet marks an overall increase in energy.) The Jovian aurorae observed by Voyager 1 stretched for almost 30,000 km across Jupiter's north pole, making them the largest such phenomena ever seen. At the same time radio emissions of a type commonly associated with the Earth's auroral regions were detected by the Plasma Wave instrument near the Io plasma torus.

Radio emissions from Jupiter

It has already been mentioned that Jupiter emits waves at radio frequencies (*see* page 10). These waves are not like the signals that carry radio programs, but resemble rather the static or "noise" that causes interference when a radio receiver is playing in the vicinity of flashes of lightning or certain electrical equipment. The radio noise reaching Earth from Jupiter is greater than from any other extraterrestrial source except the Sun and falls into three distinct types: "decametric", "decimetric" and "thermal". Each is characterized by a particular range of wavelengths, although there is some overlap between the decimetric and thermal emissions; but they are quite different in origin and sufficiently distinct in character to enable them to be considered separately.

2

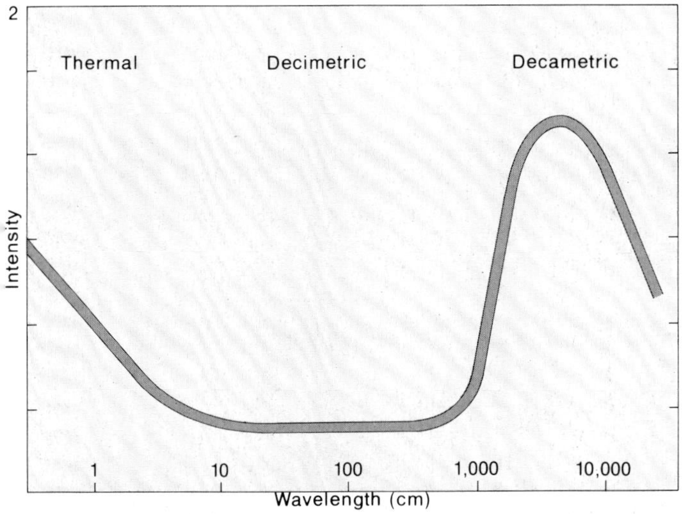

2. Radiation from Jupiter
Jupiter emits three distinct types of electromagnetic radiation at radio frequencies. This graph shows wavelength plotted against intensity. The shortest wavelengths, which are produced by thermal emissions, form the sloping part of the curve in the left-hand portion of the graph. The central, straight part of the curve represents emissions known as decimetric, which have a different character and origin from thermal radiation. There is, however, some overlap between their wavelengths. The hump in the right-hand part of the curve represents the third type of emission, decametric. Unlike the shorter wavelength radiation, decametric emissions from Jupiter are irregular.

3

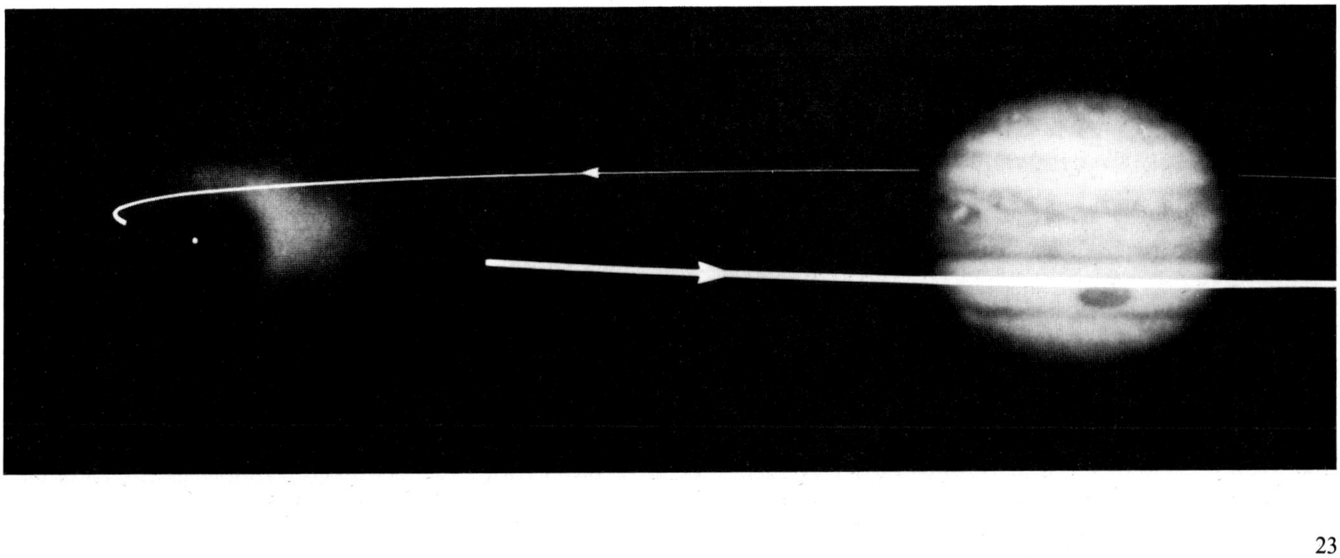

3. Io's sodium cloud
A cloud of neutral gas composed of sodium, potassium and magnesium forms a tenuous atmosphere around Io. This photograph, taken in 1977 from Table Mountain Observatory, shows the cloud stretched out along Io's orbit. A greater proportion of sodium appears to precede the satellite than that trailing behind. This image is in fact a composite, made up of an ordinary photographic image of Jupiter together with an intensified television image of the cloud. The satellite Io and the path of its orbit have subsequently been drawn onto the image.

Structure of Jupiter Radiation

Decametric emissions

The longest electromagnetic waves emitted by Jupiter have wavelengths ranging from about 7.5 m to 700 m; because the most intense activity occurs at wavelengths measured in tens of meters this type of emission is known as "decametric". Radio astronomers, however, often prefer to identify waves by their frequencies rather than by wavelength. (The two quantities are related by a simple expression: wavelength multiplied by frequency equals the speed of propagation, in this case the speed of light.) Thus Jupiter's decametric waves have frequencies in the approximate range of 40 MHz down to 425 kHz, with the peak of activity occurring between 7 and 8 MHz.

The decametric emissions are not continuous, but are characterized by strong bursts at sporadic intervals. Bursts of intense activity may last for anything between a few minutes and several hours, and are generally separated by long periods of inactivity. It was, in fact, the decametric band of the spectrum that was first observed, and the irregularity of emission and its polarization were two of the main features that led to the idea that the origin of the waves involved a magnetic field (*see* page 18). More conclusive indications came from observations of the decimetric wavelengths, described below.

Although some of the irregular decametric emissions could be created by gigantic electrical discharges such as lightning flashes in the upper atmosphere of Jupiter, they are circularly polarized (*see* Glossary) and exhibit certain regular features which are associated with the relative position of the Galilean satellite Io. It has long been known that the position of Io in its orbit profoundly influences decametric emissions, producing peak effects at certain fixed positions (*see* diagram 1), but the mechanism responsible for this effect is not yet understood.

Observations of the decametric radiation emitted by Jupiter have revealed a third system of rotation, known as System III, comparable with the System I and System II differential rotation of the visible features described on page 8. System III has a rotational period close to, but not exactly equal to, that of System II, and corresponds to 9 hr 55 min 29.710 sec or 870°.536 per day. System III is, in fact, found to represent the rotation of the magnetosphere; since the magnetic-field lines are joined to the deep interior of the planet where the field is created, System III gives the "true" period of Jupiter itself, in the sense that it behaves as if the planet possessed a solid surface like the Earth. For very accurate work, the velocities of features moving in the Jovian atmosphere are conventionally measured relative to System III, thus overcoming the lack of a fixed reference frame on Jupiter's visible surface.

Decimetric emissions

At wavelengths below 7.5 m the radiation is continuous in both time and frequency. In this band of the spectrum the shorter wavelengths corresponding to higher frequencies are predominantly produced by thermal radiation, while the lower-frequency, longer wavelengths are non-thermal in origin. These emissions, concentrated in the region of a few meters down to a few centimeters, are referred to as "decimetric".

Decimetric emissions have been observed to emanate from an area larger than Jupiter's disc. They are, moreover, distinguished from the thermal emissions by their 30 percent linear polarization (*see* Glossary) and are strongly "beamed" into the plane of the magnetic equator.

The main features of the decimetric emissions can be satisfactorily accounted for as synchrotron radiation, which occurs when electrons are accelerated along a helical path around magnetic field lines at speeds approaching the velocity of light. (Such speeds are often referred to as "relativistic", because the behavior of particles travelling at these very high velocities can only be accounted for in terms of Einstein's Theory of Relativity.) Under these conditions, electrons emit polarized radiation in narrow cones pointing along the direction of the electron's motion.

1

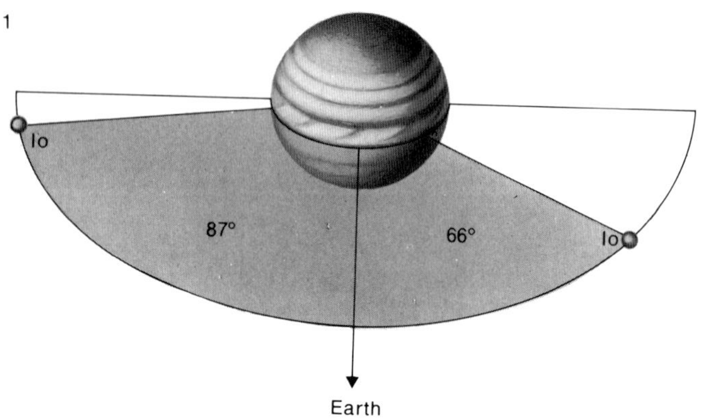

2

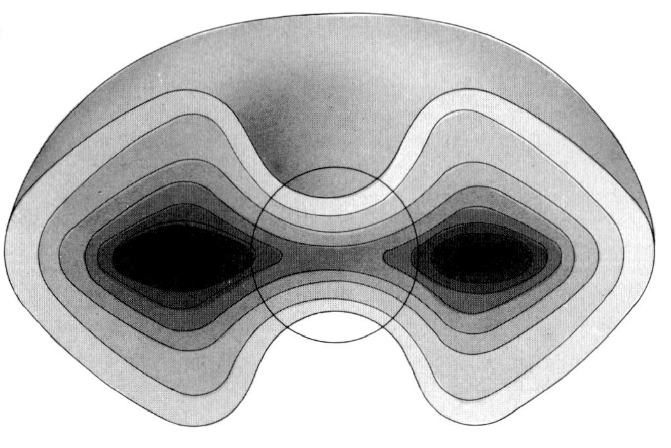

1. Decametric emissions
The irregular decametric emissions from Jupiter are known to be influenced by the position of Io relative to the Earth. The majority of stronger bursts occur when the angle between Io, Jupiter and the Earth corresponds to the values shown below.

2. Decimetric emissions
Radiation in the 10 cm band is most strongly emitted from the dark areas of the diagram.

3. Synchrotron radiation
When a fast-moving electron spirals around magnetic field lines it emits a narrow beam of strongly polarized radiation. The wavelength is dependent on the strength of the field and the velocity of the electron. Jupiter's decimetric emissions are produced by this so-called synchrotron radiation. By studying the direction of the polarization it is possible to discover the direction of the magnetic field.

3

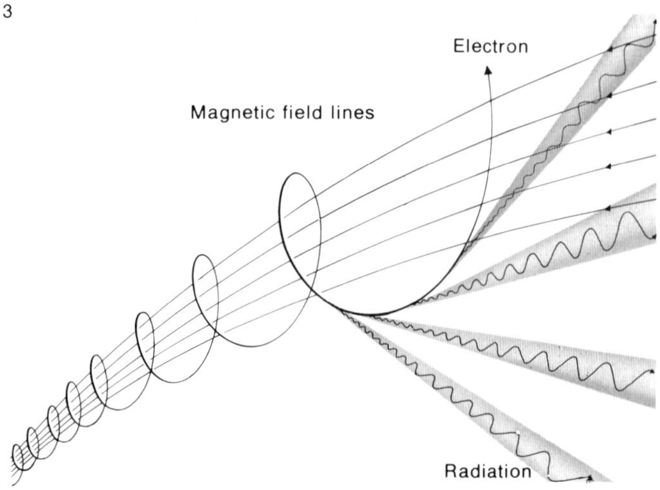

As Jupiter rotates the changing aspect of the magnetic equator as viewed from Earth causes variations in the observed intensity and polarization. The total intensity goes through two maxima and two minima per revolution, the maxima occurring when the magnetic equator is viewed edge on, and the minima when the equatorial plane is viewed from extreme positive and negative latitudes.

In addition to the linear polarization, a small amount of circular polarization has been observed. It is left-handed when Jupiter is viewed from positive magnetic latitudes, indicating that the polarity of the planet's magnetic field is opposite to that of the Earth (*see* pages 18–19).

Thermal radiation

When a body is heated it radiates energy. If a curve is drawn, plotting the intensity of radiation against wavelength, it is found to have a characteristic form which reaches its peak at a certain wavelength depending on the temperature of the body: the hotter the body, the shorter the wavelength at which the peak occurs. This corresponds with common experience: as an object is heated it starts to glow, first dark red, then yellow, and eventually white or white tinged with blue. (Physicists interpret this type of thermal radiation in terms of a theoretical model of a perfect radiator called a "black body", which is assumed to reflect no radiation whatsoever. For such a black body it is possible to predict the exact intensity-wavelength curve for any given temperature; conversely, the temperature may be determined by measuring the radiation.) Thus the thermal emissions of Jupiter at wavelengths of a few centimeters and below carry information about the temperature of the planet, and a particular wavelength is sometimes spoken of as having an "equivalent temperature".

Radiation emitted from inside Jupiter has to pass through the planet's atmosphere before it is detected. As it does so it is selectively absorbed and re-emitted by the atoms and molecules present in the atmosphere. Because the re-radiation is omnidirectional, radiation travelling in a given direction will be depleted; and since every chemical substance is known to absorb at certain characteristic wavelengths, the wavelengths at which depletion occurs provide information about the constituents of the atmosphere (*see* page 7). More detailed examination of the spectrum, continuing through the thermal radio wavelengths to the infrared and visible wavelengths, can also yield information about the changes of temperature and pressure with altitude in the Jovian atmosphere. If, for example, the temperature of the atmosphere increases instead of decreases with altitude, the gases at higher levels will appear to be emitting rather than absorbing radiation; their characteristic "spectral lines" will show up as stronger rather than weaker emissions at particular wavelengths.

Thermal maps

Analysis of the spectrum of Jupiter, particularly of the thermal radiation at infrared wavelengths, has made it possible to build up a picture of the structure of the Jovian atmosphere. Breaks in Jupiter's thick cloud layers allow measurements to be taken over a range of depths down to layers where the pressure is several Earth atmospheres. Infrared radiation penetrates haze more easily than visible radiation, and observations at a wavelength of 5 μm are ideal for studying the deepest Jovian cloud layers, because there is negligible absorption of this wavelength by the main chemical constituents of the atmosphere (ammonia, methane and hydrogen). Other infrared wavelengths are known to be most strongly emitted at certain pressures and can therefore be used to produce "thermal maps" of the varying structure of the Jovian clouds. The atmosphere is too opaque to permit much information to be obtained about layers deep below the clouds, but the available data on the upper regions indicate a structure that may usefully be compared with that of the terrestrial atmosphere. Jupiter's characteristic structure of alternating belts and zones is clearly evident in thermal maps such as the two shown on this page (*see* diagram 5).

4. 5 μm maps
An Earth-based image of Jupiter at a wavelength of 5 μm (**A**) reveals "hot" regions corresponding with the dark belts seen in the picture made by Voyager 1 (**B**) about an hour later. These hot 5 μm areas indicate the deeper levels of Jupiter's atmosphere, where there is little or no overlying cloud. At 5 μm the GRS appears cooler than its surroundings.

5. Thermal maps
Jupiter is shown at two different infrared wavelengths, 44.2 μm (**A**) and 16.6 μm (**B**). Map A thus depicts the temperature structure of the atmosphere at the level of the cloud tops, while map B represents a higher altitude corresponding with the tropopause (the region between the troposphere and the stratosphere). The cool region on map B at about 23°S, 105°W is associated with the GRS; no corresponding feature is apparent on map A.

4A

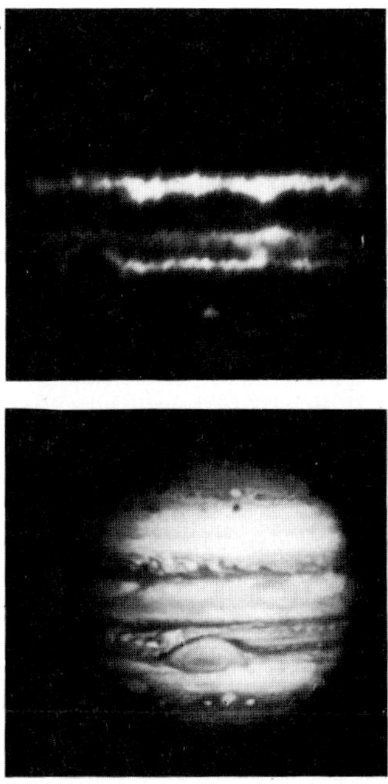

B

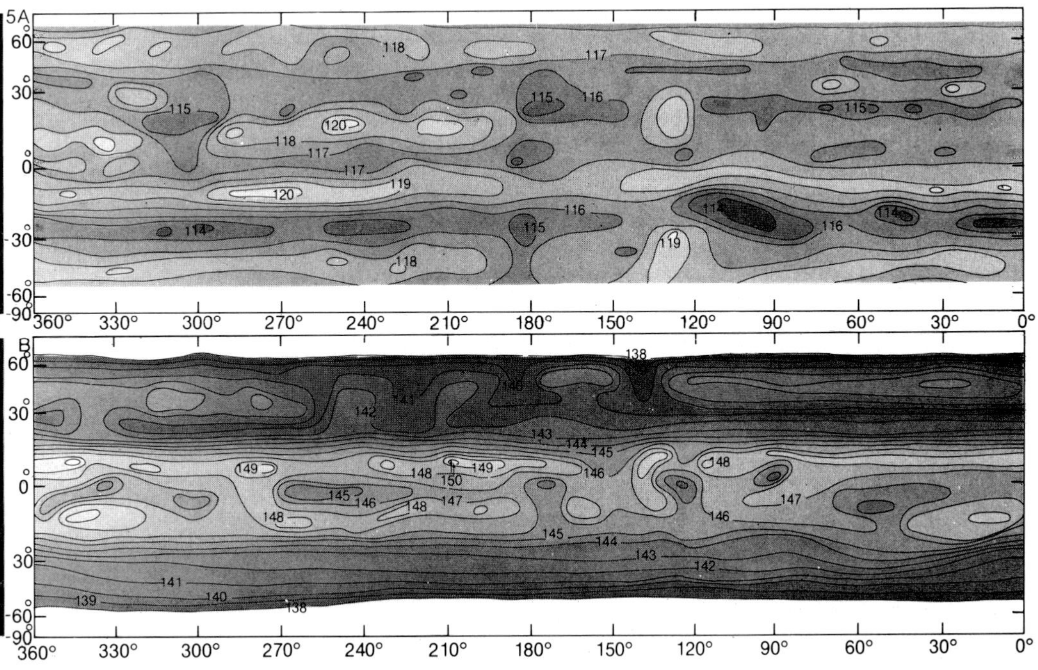

Structure of Jupiter Atmosphere

The structure of the atmosphere
The Earth's uncharged or "neutral" atmosphere is divided into several distinct regions. The innermost layer is called the troposphere; above it lies the stratosphere, followed by the mesosphere and, then, the thermosphere. These layers are primarily distinguished by the way in which temperature varies with altitude as different influences make themselves felt. The thermosphere has an important, electrically charged region within it called the "ionosphere". Finally comes the exosphere, which has no definite boundary, but whose density falls off with altitude until it is no greater than that of the interplanetary medium.

The troposphere
On Earth the temperature above the planet's surface decreases with increasing altitude until it reaches a minimum at a height of approximately 15 km; this region is the troposphere. One of the main constituents of the troposphere is water vapor, and it is here that the weather processes, in particular the formation of clouds, take place. The dominant mechanism by which heat is transported vertically is convection. Air, warmed by the surface of the Earth, rises and is cooled higher up, while cool air is drawn downwards to replace it and is warmed in turn. The equivalent region on Jupiter is bounded at the upper point by a temperature minimum of about 105 K, which occurs when the pressure is about 0.1 Earth atmospheres. As on Earth the exact value of the temperature minimum varies with different locations on Jupiter's globe. As observed by Voyagers 1 and 2, the coldest value occurred over the Great Red Spot and in the zones. Below the point of temperature minimum the temperature generally increases with depth, following closely the "adiabatic" gradient of about 2° km^{-1}. (This is the value for a well-mixed atmosphere, and implies that the region, like the Earth's troposphere, is stirred by convective processes.)

The stratosphere and mesosphere
By analogy with Earth, the temperature minimum in Jupiter's atmosphere marks the transition between the troposphere and the stratosphere, a region in which the temperature is controlled largely by radiation processes. On Earth the change in temperature gradient (the rate of change of temperature with height) is due to the heating of the upper atmosphere by a layer of ozone, an "allotropic" form of oxygen (*see* Glossary) which absorbs ultraviolet radiation from the Sun. On Jupiter the increase in temperature is due to the heating of the atmosphere by methane, which appears to play a similar role to the Earth's ozone. Dust, possibly from the Jovian ring system (*see* pages 56–57), may also play a part in heating Jupiter's stratosphere.

The temperature gradient of the Earth's atmosphere changes again at an altitude of about 30 km, when temperature once again decreases with height for a further 30 to 40 km; this region is known as the mesosphere. However, not enough is known about Jupiter's atmosphere to describe in any detail the region corresponding to the terrestrial mesosphere.

The ionosphere
In the upper regions of the atmosphere where the density is very low, the electrical conductivity increases. This part of the atmosphere, within the thermosphere, is another distinct region known as the ionosphere. The name is derived from the high proportion of ionized atoms and molecules contained in this region, and the conductivity is due to the consequent presence of free electrons. Jupiter's ionosphere, extending for more than 3,000 km above the visible surface, is comparable with that of the Earth, which begins at altitudes of around 80 km to 500 km above the surface. The atmosphere here is highly rarefied, and the air molecules are easily ionized by energetic radiation from the Sun with wavelengths of 1,000 Å or less (in other words, ultraviolet radiation). By this process, energy from the Sun in the ultraviolet portion of the spectrum is absorbed by the atoms and molecules in upper regions of the atmosphere, causing them to eject an electron and leave behind a positively charged ion. The Earth has, in fact, several distinct ionized layers, extending upwards as high as 300 km. These layers have properties which make them important in radio communications; they reflect back certain radio frequencies transmitted from the Earth, thus making it possible to send signals around the globe. At the same time, however, they also reflect radio waves back out into space, and therefore obstruct certain types of astronomical research.

The extreme ultraviolet radiation that causes photoionization originates in the upper chromosphere and corona of the Sun. Changes in these regions of the solar atmosphere therefore cause detectable variations in the density of electrons in the ionosphere.

Jupiter's ionosphere is a huge, highly structured region, whose principal constituent is ionized hydrogen (H^+ ions) produced by the same process of photoionization as in the Earth's ionosphere. The structure of this region changed appreciably between the Pioneer and Voyager encounters. The ionosphere seen by the Voyager had a much greater vertical extent than had been observed by the Pioneer spacecraft four and a half years earlier, as well as exhibiting greater diurnal variation. There is no doubt that these differences result from increased solar activity (*see* page 20).

With Jupiter the energetic particles in the magnetosphere are likely to provide an important secondary source of ionization.

1

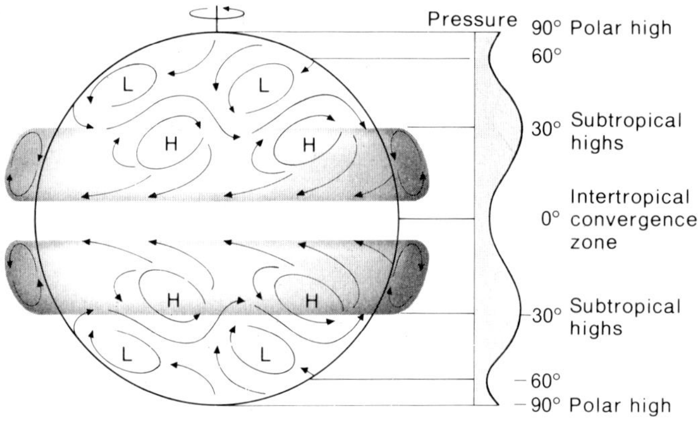

1. The Earth's circulation
A large temperature gradient between the equator and the poles dominates the Earth's circulation, producing a longitudinal wave travelling from west to east.

2. The Earth's atmosphere
The structure of the Earth's atmosphere is shown in this graph. Changes in the gradient of the curve indicate the important boundaries.

2

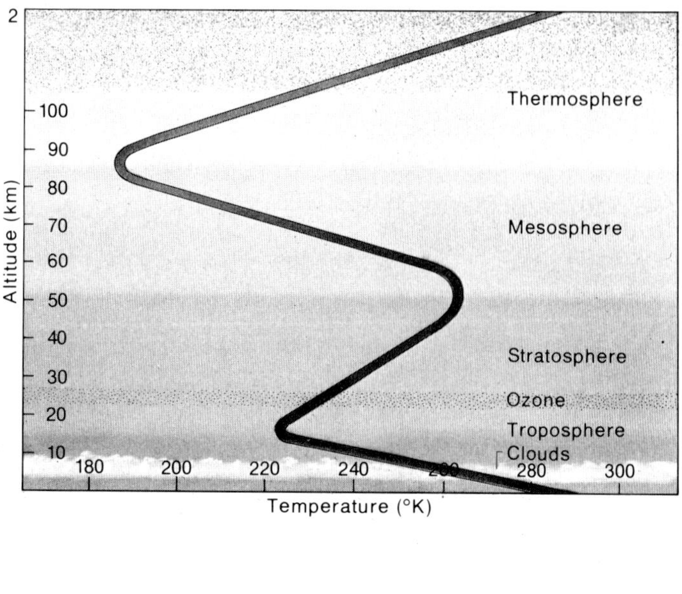

Their contribution to the production of the ionosphere would certainly account for the strong latitudinal variations observed by the Pioneer and Voyager spacecraft, and for the differences in ionosphere profiles.

Circulation of the atmosphere

It is, then, in the lower levels of the atmosphere that the meteorological processes which produce the visible features of the planet take place. At first glance Jupiter's appearance suggests that its weather systems are quite different from those found on Earth or, indeed, on any of the other terrestrial planets such as Mars or Venus. Jupiter is a completely fluid planet (as least, if it has a solid core, it is only a very small proportion of the total planetary radius) and thus has no solid surface equivalent to that of the Earth affecting the atmospheric motions. There are also fundamental differences in the driving mechanisms of Jupiter's weather system. For example, on Jupiter the temperature at the poles is virtually the same as the temperature at the equator, whereas on Earth the poles are much colder than the equatorial regions, and the transfer of heat outwards from the equator plays an important role in the Earth's meteorology; thus a longitudinal wave is set up moving in a westerly direction, transporting the excess heat to the poles.

Another important difference between the driving mechanisms on Earth and Jupiter affecting their weather is found in their sources of heat. The primary source of energy for the Earth's weather system is solar radiation; by contrast, Jupiter has an internal source of heat energy, but receives only relatively weak incident sunlight because of its distance from the Sun. Finally, the rapid 10 hr rotation of Jupiter plays an important part in shaping the planet's clouds.

Nevertheless, many of the individual features on Jupiter's visible surface may be interpreted through analogies with familiar terrestrial weather-effects, and it may indeed be that the whole of Jupiter's weather system is no more than a multiple form of processes that are relatively well-understood from the Earth's atmosphere. For example, on Earth air will flow outwards from a region of high pressure, producing an anticyclone which spirals in a clockwise direction in the northern hemisphere or anticlockwise in the southern hemisphere; for low-pressure regions, the direction of flow is inwards to produce a cyclone, spiralling in the opposite direction to an anticyclone in either hemisphere. (The direction of flow depends on the Coriolis forces—*see* Glossary.) On Jupiter the regions of high and low pressure are stretched around the planet by the rapid rotation of the body, producing the characteristic bright zones and dark belts respectively. There is a temperature difference between these regions amounting to 3 K or less, and although this difference is only slight by terrestrial standards, it may, coupled with the rapid rotation, energize the circulation in such a way as to create the observed pattern of alternating easterly and westerly jets. Thus Jupiter's belts and zones may simply be large-scale wave systems, extending to considerable depths throughout the meteorologically active regions, but produced by the same fundamental forces that operate in the Earth's atmosphere.

In recent years considerable progress has been made in studies of the Earth's meteorology with numerical models which represent in detail the physical process and surface conditions over the planet. One such model has been extended to the conditions of the Jovian atmosphere. With this type of approach it is possible to vary different parameters so as to examine the response of the atmosphere. The results, generated by a computer, closely resemble the large-scale features of the Jovian atmosphere, such as the belts, zones and the large-scale spots. It would seem that instabilities in the atmosphere energize the circulation of Jupiter, creating large-scale planetary waves resembling the belts and zones, while the jets result from the turbulent interaction of propagating waves. The absence of any solid surface on Jupiter probably contributes to the distinctive zonal appearance.

3

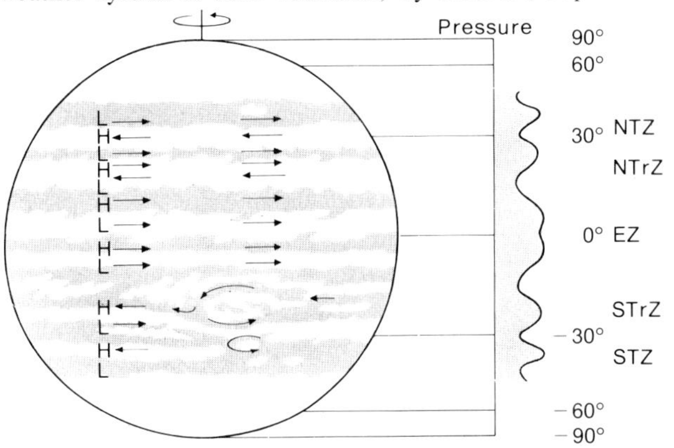

3. Jupiter's circulation
On Jupiter there is little flow between latitudes. Instead there are alternating regions of high and low pressure, with high-speed winds between the bands.

4. Jupiter's atmosphere
The basic structure of Jupiter's atmosphere resembles that of the Earth, at least at tropospheric levels. However, knowledge of higher levels is still incomplete.

5. Computer models
Mathematical models representing the Earth's atmosphere have been programmed with data for Jupiter. The results shown here depict three stages of the model's development. The image gradually transforms itself until flow-patterns evolve which closely resemble the alternating zones and belts found on Jupiter. Forms resembling the GRS also occur.

4

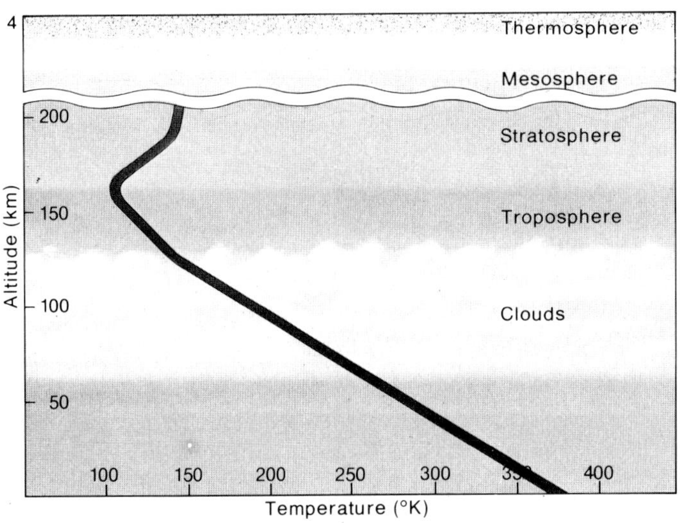

5

Structure of Jupiter Cloud Morphology

Cloud shapes

The changing shapes of Jupiter's cloud surface have been observed from Earth almost continuously over the past century, and considerable variations have been seen to occur, sometimes over very short periods. In the equatorial region marked changes sometimes take place within a length of time less than the planet's period of rotation. In the four and a half years between the Pioneer and the Voyager flybys, Jupiter's appearance had become considerably more turbulent and dramatic. In the broadest white zone, for example, violent motions can be seen in the Voyager photographs that are not apparent in the earlier photographs. Other features, however, remain stable over long periods; two plume-like clouds trailing behind (to the west of) small bright centers observed by the Pioneer spacecraft have been traced back a whole Jovian year (almost 12 Earth years), while features like the Great Red Spot and the white ovals have lifetimes of hundreds of years. From the Voyager studies it was found that all the easterly jets are unstable, while the corresponding westerlies remain stable.

The northern hemisphere

Generally speaking Jupiter's northern hemisphere is distinguished from its southern counterpart by its lack of large-scale structures comparable to the Great Red Spot, although it does exhibit smaller features which appear to be comparable in basic morphology. This asymmetry between north and south is one of the planet's most striking characteristics.

At the time of the Voyager 1 encounter, 13 plume-like features were observed between the equator and approximately 10° N, in the Equatorial Zone. Then, a few months later, only 11 plumes were seen (*see* pages 38–39). This sudden change is caused by rapid upward motions in convective cloud systems which have horizontal scales of about 2,000 km and contain individual puffy elements about 100 to 200 km in size. The energy for these streams must be related to the condensation processing occurring in layers of water-cloud beneath the visible clouds. Morphologically, these smaller clouds resemble terrestrial cumulus, and contrast sharply with the diffuse, filamentary features found in many other regions on Jupiter. The flow characteristics of the plumes resemble some of the properties of clouds found in the Earth's tropics, in the Inter-Tropical Convergence Zone (ITCZ), where the crests of a travelling wave trigger convective activity in the atmosphere.

The plumes move in a westerly direction in a strong equatorial jet whose velocity ranges from 100 to 150 $m\,s^{-1}$. Surprisingly, these plumes are only found along the northern boundary of the Equatorial Zone and have no counterparts in the south. However, the flow that probably triggers the plumes requires a convergence of fluid beneath the visible surface, and in the southern part of the Equatorial Zone such a flow might well be disrupted by the presence

1

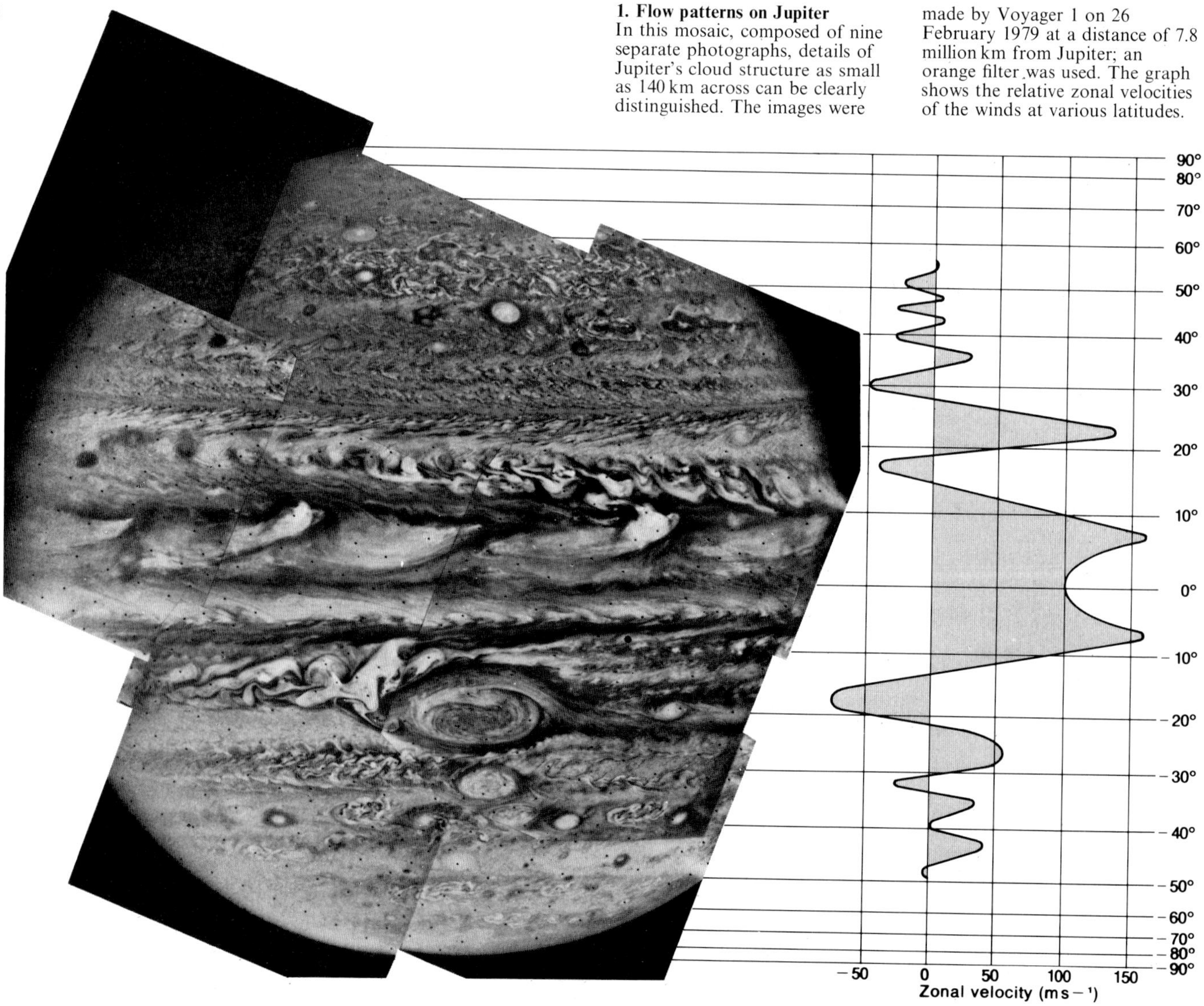

1. Flow patterns on Jupiter
In this mosaic, composed of nine separate photographs, details of Jupiter's cloud structure as small as 140 km across can be clearly distinguished. The images were made by Voyager 1 on 26 February 1979 at a distance of 7.8 million km from Jupiter; an orange filter was used. The graph shows the relative zonal velocities of the winds at various latitudes.

of the giant cloud systems such as the Great Red Spot and the white ovals. Thus these large features may account for the asymmetry between north and south in this region.

Further north, in the North Equatorial Belt (latitudes 9° to 18°N), cloud patterns vary considerably with longitude, with dark brown cloud shapes at irregular intervals. These structures generally have sharp boundaries and sometimes show linear striations in their interiors. The clouds seem to rotate around them cyclonically, indicating that these dark regions represent low-pressure systems.

As the brown clouds moved towards the west with the prevailing current, significant changes occurred in the period between the two Voyager encounters. The most westward feature almost disappeared, and a small but important intrusion of white appeared in the largest of the dark clouds (*see* illustration). It is clear that the different colored clouds are stratified into layers, with the brown clouds situated deeper than the bright layers (*see* page 32).

The North Equatorial Belt also marks the highest 5 μm emissions. It is therefore apparent that the visible features in this region are the deepest that can be seen in Jupiter's atmosphere, where the temperatures and pressures are high. (The cloud tops radiate 5 μm waves with an equivalent temperature of about 140 K, as compared with the maximum of 258 K found at the deepest observable levels. Even the highest temperatures, however, are lower than those found on the surface of the Earth on a warm day.)

On the southern edge of the North Temperate Belt (latitudes 20° to 23°N) the clouds take on a tilted linear pattern associated with the opposing currents at 18° to 23°N. These features tilt north-east and form chevron shapes with the rows of puffy cloud tilted in the opposite direction further to the north. The apex of the chevron pattern lies in the strongest part of the westerly jet, which had a speed of 150 m s^{-1} during the Voyager encounter. In 1970 these features were found to move at a speed of 163 m s^{-1}, the highest known zonal velocity in the Jovian atmosphere.

The high-speed jet appears as a thin brown line dividing the broad and apparently featureless North Tropical Zone. In the northern boundary of this zone, at 35°N, a row of dark clouds can be seen. The dark areas surrounding the bright centers are associated with high 5 μm radiation. These small circular features measure about 3,000 km in diameter. They are composed of dark rings with white cones, and rotate anticyclonically. On one occasion there was a most extraordinary interaction, when the poleward member of a pair overtook the other and combined with it (*see* illustration). The combined features tumbled for a while and then ejected a streamer to the west, towards the equator. The new, combined spot then continued to proceed to the east.

Between the two Voyager encounters the spacing between the spots in this region altered considerably, indicating strong motions of these spots relative to one another. The mean zonal speed is about 26 m s^{-1}.

Further north the alternating pattern of belts and zones breaks down. In regions north of 35° long recirculating currents are common features, though occasionally there are large cloud systems such as the one at 45°N, 70°W, which appears to be more than 12,000 km in diameter, with a spiral structure in its interior.

2. Temperature graphs
These graphs depict the temperature of Jupiter's atmosphere at the latitude of the plumes. The lower of the two lines represents the temperatures at a height of 15 km above the cloud tops, while the upper line indicates higher temperatures at the level of the cloud tops.

3. Plumes
This photograph shows a detail of two of the plume-like features at latitudes 0° to 10°N. Some of these features appear to remain stable over relatively long periods, although the number of plumes seen at any one time varies.

4. Dark cloud
A wisp of light cloud protrudes over one of the long dark clouds in the NEB, clearly demonstrating that the dark region is situated at a lower level. The image was made by Voyager 2 on 6 July 1979 from a distance of 3.2 million km.

5. Chevron patterns
A high-speed jet separates the NTrZ from the NTZ. Linear features in the two zones tilt towards one another, forming V-shaped patterns where they meet.

2
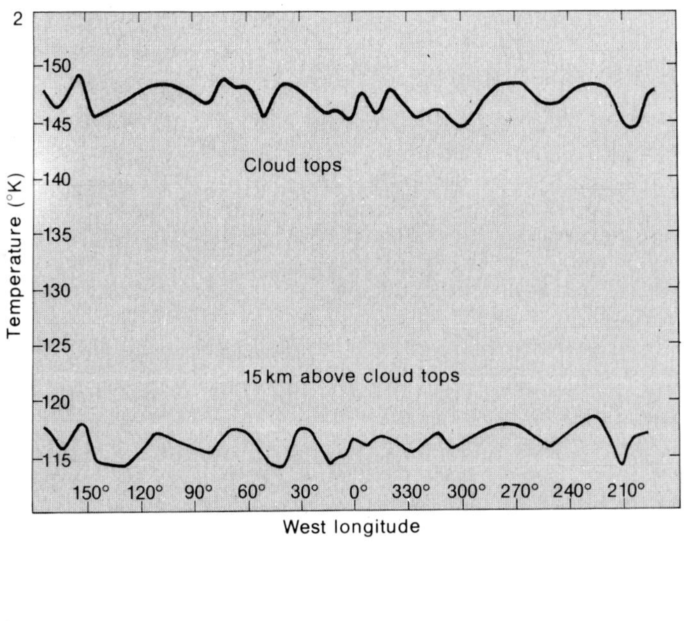

3
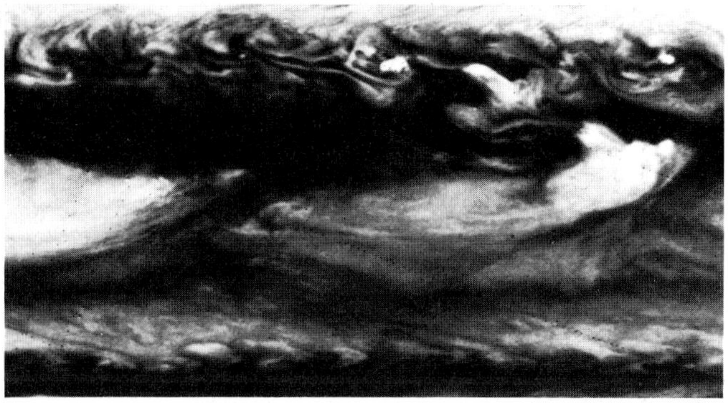

4
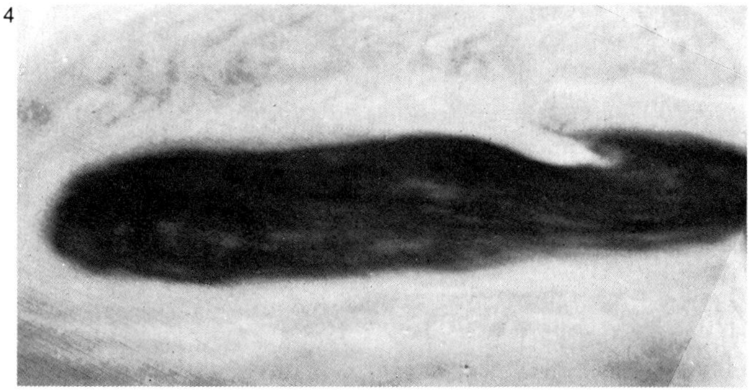

5
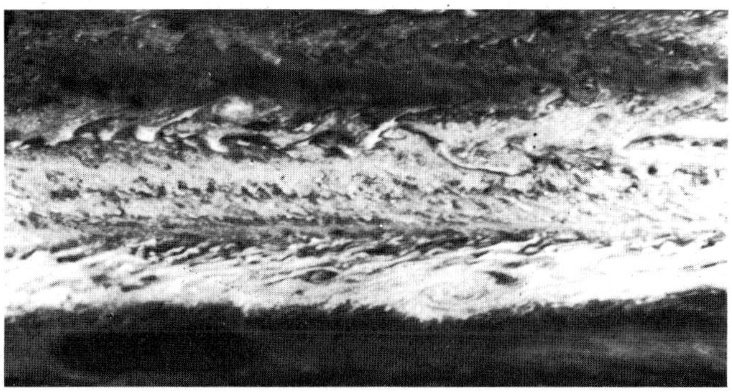

Structure of Jupiter The Great Red Spot

The southern hemisphere

The appearance of Jupiter's southern hemisphere is dominated by the spectacular Great Red Spot and the three smaller, though similar, white ovals. A row of spots can also be seen at 41°S, and these features may also be related in structure to the large spots. The history of observations of the Great Red Spot and early theories about its nature have already been described (*see* pages 10–13), but the information gained from the satellite missions now places major constraints on theories about it.

The Great Red Spot is not a fixed feature. During the period covered by the photograph (*see* illustration) it drifted westward at a rate of approximately 0°.5 per day. In addition it has been found to oscillate relative to the north–south axis, with an average amplitude of 1,800 km. The period of oscillation is 89.85 ± 0.1 days. The Great Red Spot is not unique in color; very small red spots have been observed from time to time in the northern hemisphere of the planet. The most prominent was seen during the Pioneer 10 flyby.

Associated with the Great Red Spot is a disturbed region to the west of the feature, in which sudden brightening frequently breaks out. Similar features seem to be characteristic of all the southern hemisphere spots (*see* illustration). In fact, the white ovals and the spots at 41°S resemble the Great Red Spot in several ways. They all rotate in an anticyclonic manner, and many have associated cyclonic circulating currents to the east and cyclonic "wake-like" regions to the west. (The resemblance to a wake is misleading. Movement of features in this region is eastward towards the Great Red Spot rather than westward away from it; such features then appear to be blocked by the Great Red Spot. An exception occurred when some light material crossed the South Equatorial Belt and then centered in the rapid westerly current in the Equatorial Zone.)

At the time of the Voyager 1 encounter, small spots were seen interacting with the Great Red Spot. Cloud vortices, moving in an easterly direction towards the Great Red Spot at a speed of about 55 $m s^{-1}$, were deflected northward at the eastern cusp, and entered the westerly current flowing on the north side of the Great Red Spot. Some of the cloud spots circulated around the Great Red Spot with a period of six days, approximately half the time taken 20 years ago when similar interactions were first observed.

As the spots rotate around the Great Red Spot, their shapes become distorted by the opposing directions of flow between the regions through which they are moving. In the interval between the two Voyager encounters, a cloud structure started to develop to the east of the Great Red Spot, forming a barrier to the flow (*see* illustration). While cloud vortices continued to approach the Great Red Spot at about 55 $m s^{-1}$, the presence of this barrier forced them to recirculate in the direction from which they originally came. Considerable shear was set up in this region, causing the cloud vortices to become distorted; as many as four or five such vortices were seen in this recirculating pattern adjacent to the Great Red Spot, which itself continued to rotate just as before.

The appearance of this cloud structure east of the Great Red Spot is highly significant. It may indicate a transition in the appearance of the planet, possibly corresponding to the change from placid to turbulent that took place between the Pioneer and Voyager encounters. This change of appearance may be part of a climatic cycle in the Jovian atmosphere.

Nature and origin of the Great Red Spot

Infrared observations have provided some important clues to understanding the Great Red Spot. At 5 μm wavelength, the Spot is much colder than its surroundings, suggesting that it is an elevated high-pressure region; this idea is consistent with its anticyclonic rotation. Temperature maps of the Great Red Spot and of the white

1. The Great Red Spot
A time-lapse sequence of photographs showing the GRS after every alternate rotation of Jupiter reveals the flow of material around the GRS. This particular sequence was produced by Voyager 2.

1

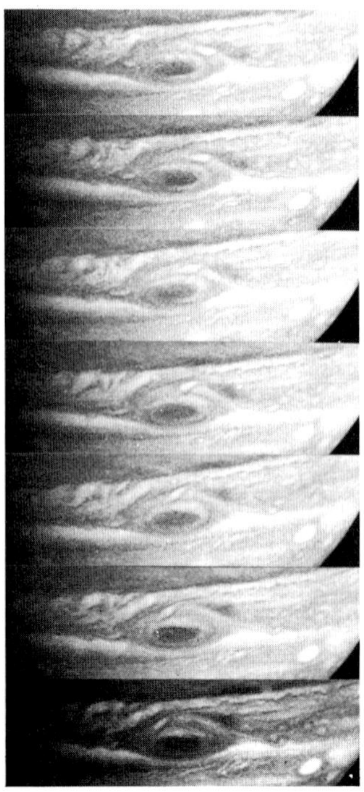

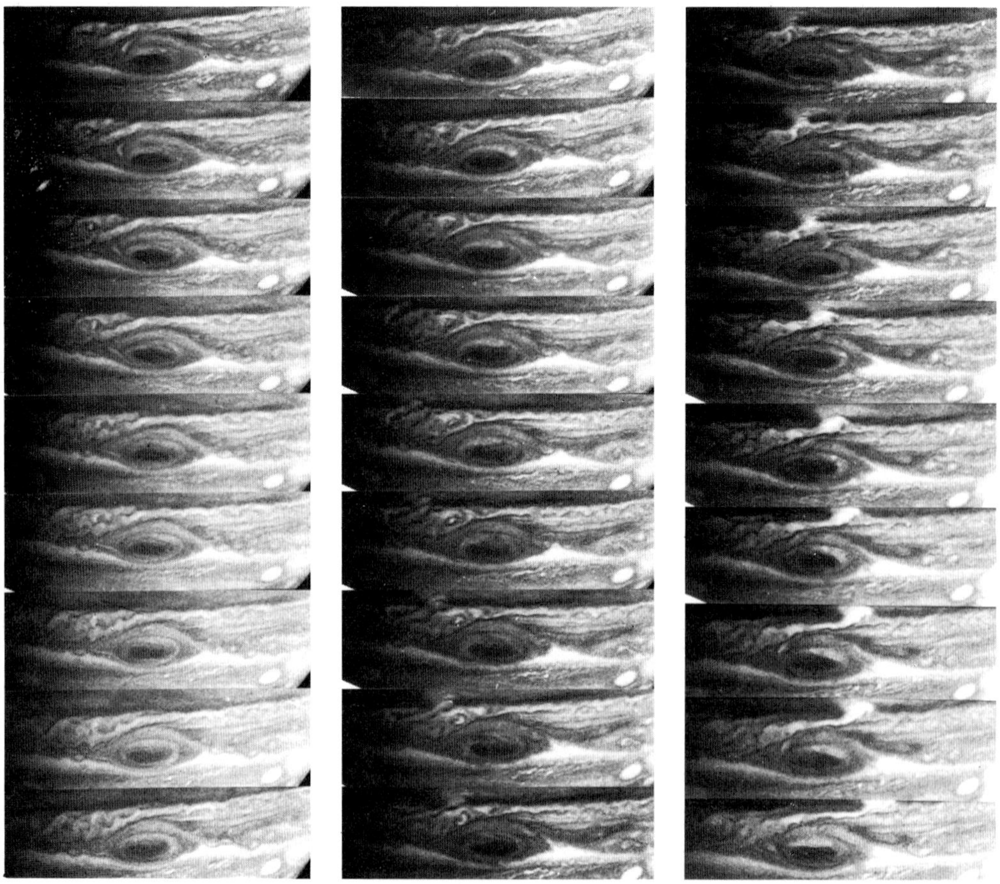

ovals show them to have cold regions above their centers, consistent with a slow upward motion of gas at a rate of a few millimeters per second, cooling "adiabatically" as it rises. ("Adiabatic cooling" takes place when a gas does not undergo a net loss of heat energy, but decreases in temperature only as a result of expansion as its pressure drops; such cooling takes place in a well-mixed atmosphere.) In a typical convective system on the Earth, upward motions are much faster, in the region of a few centimeters per second. If material is flowing upwards in the Great Red Spot and the ovals, there must be a corresponding convergence of material deep beneath the clouds to replace the material flowing up. The new material would slowly spiral up into the circulating system and then flow outwards at the level of the cloud tops.

Perhaps the best clue to understanding the nature and origin of the Great Red Spot comes from the white ovals, which were observed at their formation in 1939. At that time dark features were seen in the South Temperate Zone, stretching all the way around the planet. Gradually these dark regions extended until three large, white oval features formed, each nearly 100,000 km long. In the subsequent 40 years these white ovals have slowly diminished in size, and are now only 11,000 km long. At the time of the Voyager 1 encounter they were positioned at longitudes 5°, 85° and 170°W.

It would seem, therefore, that the white ovals may well disappear in the very near future. They may simply be large-scale gyres formed out of the zone in which they are situated. If they are merely temporary, there would be no need to find a sustained driving mechanism in order to explain them. The Great Red Spot may have originated in a similar fashion, and may also be a transient feature in the Jovian atmosphere, although it appears to be too long-lived for this explanation to be entirely satisfactory. However, like the white ovals, it is known to be shrinking, and is now half the length it used to be a century ago.

There have been, however, several other theories, such as the idea of a "Taylor column" described earlier (*see* page 12). One interesting recent suggestion is that the Great Red Spot is a kind of "soliton" or "solitary wave", an isolated permanent wave which can flow between two layers of fluid when there is a velocity difference between the layers. The wave takes its energy from the velocity gradient. Solitons have been analyzed from laboratory models, and a simple example is illustrated (*see* diagram 2). When solitons interact there is a phase shift that looks like an acceleration and the waves then emerge with identical forms as before. Such interactions have been compared with the interactions between the Great Red Spot and the South Tropical Disturbance, which were first observed in 1902 (*see* pages 10–11), and both the observed flows and the manner in which the features re-established themselves after interacting have a marked resemblance to the behavior of solitary waves. However, although several alternative models are able to reproduce the streamlined patterns, the uniqueness of the feature is difficult to establish. Perhaps the meteorology of Jupiter is in fact hybrid, a combination of several of the suggestions that have so far been advanced. For example, the computer model of the atmosphere (*see* page 27) would indicate that the Great Red Spot is simply a gyre; but the interaction of the Great Red Spot with the cloud vortices, observed by Voyager, and with the South Tropical Disturbance still needs to be explained. These aspects of the flows are well represented by solitary wave interactions. Perhaps some modified form of solitary wave may occur under the special conditions of Jupiter's environment. A complete explanation of the Great Red Spot would have to demonstrate such an effect within the context of a general mathematical model of the atmosphere as a whole, such as the one used in computer simulations (*see* page 27). At present the soliton idea seems to be the most promising line of investigation.

2

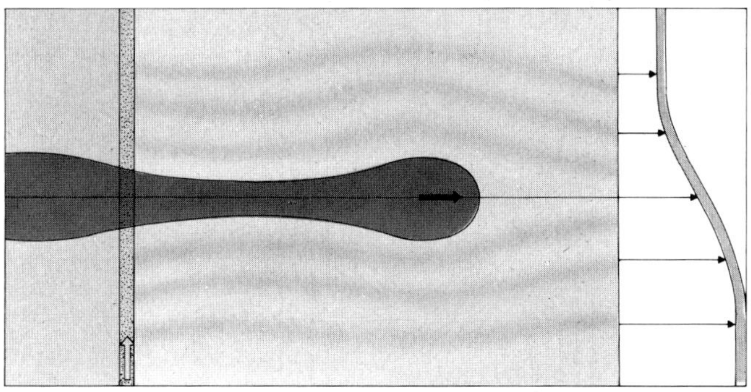

2. Solitary waves
The diagram shows a propagating solitary wave in a laboratory demonstration. When the barrier at the left is removed, the lower-density fluid forms a solitary wave which travels straight through the main tank.

3. Structure of the GRS
The GRS is believed to be an elevated high-pressure region. It rotates anticyclonically (anticlockwise), presumably drawing material upwards from some level below the visible surface of the cloud layers. This material would then flow back downwards, probably at the edges of the GRS. The exact details of these flows, however, are not known at present. At first sight the GRS and the white ovals appear to resemble terrestrial storm systems such as hurricanes, but this impression is misleading for the reasons given.

4. Terrestrial hurricane
On Earth storm clouds are driven by energy derived from a flow of air over a warm surface, such as an ocean. The hurricane is cyclonic at sea-level and weakly anticyclonic at high altitudes, and has a characteristic "eye" in the center. Air is drawn upwards and cools in the upper atmosphere, spreading outwards as it does so. The rotation of the Earth induces the clouds to form into spiral shapes. By contrast, the GRS and white ovals exhibit no central "eye" and no comparable spiral patterns. Moreover, if the Jovian features were storms, it would be necessary to explain why the whole surface of the planet was not covered with similar spots.

3

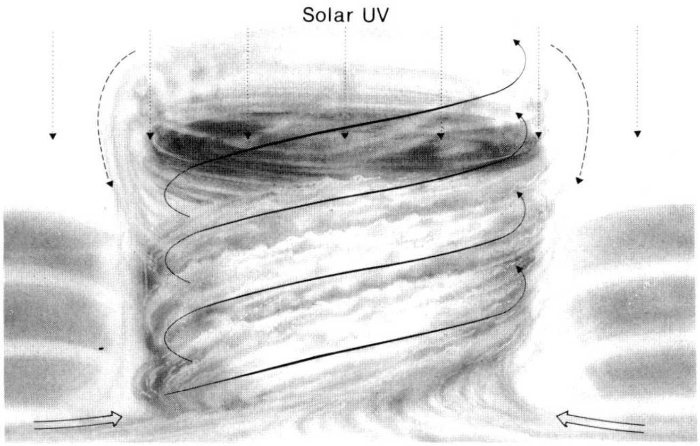

4

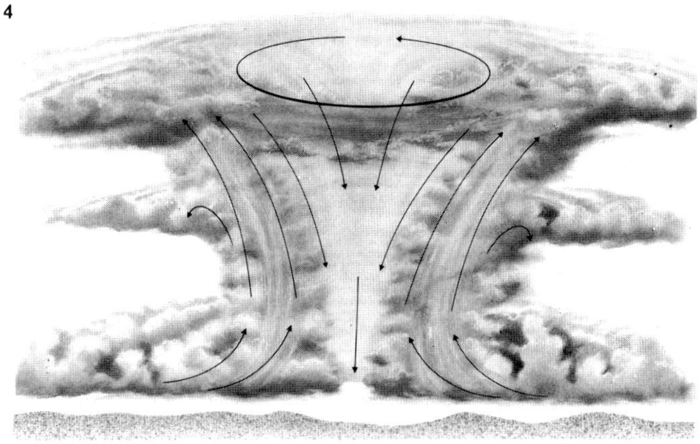

Structure of Jupiter Cloud Layers

The origin of the colors

One of the most fundamental problems concerning Jupiter is to explain the origin of the colors in its atmosphere. The Earth's clouds, composed of water and particles of water-ice, are white; but the colorful appearance of Jupiter's clouds provides vivid evidence of the differences in the chemical composition of the two planets. Ammonia, which is known to be a constituent of the Jovian atmosphere, would condense into a layer of cloud at some particular level in the troposphere; it is believed that ammonia clouds mark the top of the bright zones. The dark belts are deeper layers of cloud which are not obscured by the white veil of ammonia cloud and are therefore visible.

Beneath the cirrus-like clouds of ammonia, other layers would be expected to form, possibly involving ammonium hydrosulphide (NH_4SH), water-ice and ammonia solution. As temperatures decrease with height, different colors are seen, probably from the products of complicated chemical reactions. At mid-altitudes orange and yellow colors can be seen. It is noticeable, however, that even at the highest resolution available the colors of these clouds do not appear to merge together, but retain their individual identities.

Jupiter's atmosphere is known to include small but important amounts of hydrocarbons, such as methane and other exotic materials (*see* page 7). It is bathed in ultraviolet sunlight, which is a strong source of energy for chemical reactions. There are also violent lightning flashes which, like terrestrial storms, are associated with regions of strong convective activity. The Jovian lightning bolts are comparable in strength with the superbolts found near the cloud tops on the Earth. These storms extend throughout the clouds, and provide a second important source of energy to affect the complicated atmospheric chemistry at a variety of levels.

The colorful clouds which lie in broad longitudinal bands are thought to be created from a mixture of methane, ammonia and, probably, sulphur. The molecules are broken up by ultraviolet sunlight and violent lightning, and then stirred up by the dynamic weather patterns. A possible key to understanding the diversity of colors would be the identification of a sulphur-bearing substance, since sulphur and certain of its compounds produce a range of colors from yellow to brown or black depending on the temperature. One suitable candidate is hydrogen sulphide (H_2S), from which might form hydrogen polysulphide (H_XS_Y) or ammonium polysulphide ($(NH_4)_XS_Y$). These chemicals are capable of producing the appropriate colors.

Somehow the color of the Great Red Spot and the smaller red spots must also be accounted for within the same interpretation. The red color may be related to the presence of phosphine in the atmosphere; the action of sunlight would ultimately produce red phosphorus where phosphine is brought to the surface. For this to be possible, the red spots would have to penetrate more deeply into the atmosphere than the white ovals so as to reach the levels where phosphine exists; moreover, there is a possibility that hydrocarbons such as acetylene and ethane could act as scavengers to the phosphine reactions, reducing the amount of phosphorus produced. Consequently, considerable variations would be expected in the concentrations of these hydrocarbons over the surface of Jupiter, and this may help to explain the variety of colors and shades that are found.

More speculative suggestions have also been made, involving the idea that the observed colors may be related to organic molecules: this conjures up a fanciful image of Jupiter as a giant chemical reactor producing the same basic materials from which primitive life developed on Earth. It would seem that the Jovian atmosphere might conceivably be of interest to biologists and biochemists as well as to astronomers and meteorologists.

Cloud layers

The bands of colors on Jupiter's visible surface distinguish different levels in the planet's cloud structure. The highest clouds, which form the bright zones, consist of ammonia crystals at a pressure of about one atmosphere. About 100 km below these, ammonium hydrosulphide crystals are expected, forming a dark cloud layer; this is visible as the dark belts where there are gaps in the upper clouds. Below these there is believed to be another layer, this time composed of water-ice crystals, having a bluish appearance. A longitudinal wave is thought to travel through these layers. The hypothetical movement is illustrated by the arrows in the diagram below.

Color Plates

Jupiter, Io, Europa and Callisto
Voyager 1
5 February 1979

Jupiter
Mt. Lemmon Observatory
10 December 1975

Jupiter
Pioneer 10 (top left)
2 December 1973

Jupiter
Pioneer 11 (top right)
6 December 1974

Jupiter
Voyager 1 (bottom left)
24 January 1979

Jupiter
Voyager 2 (bottom right)
9 May 1979

North Temperate Zone
Voyager 1 (top)
2 March 1979

Equatorial Zone
Voyager 2 (bottom)
28 June 1979

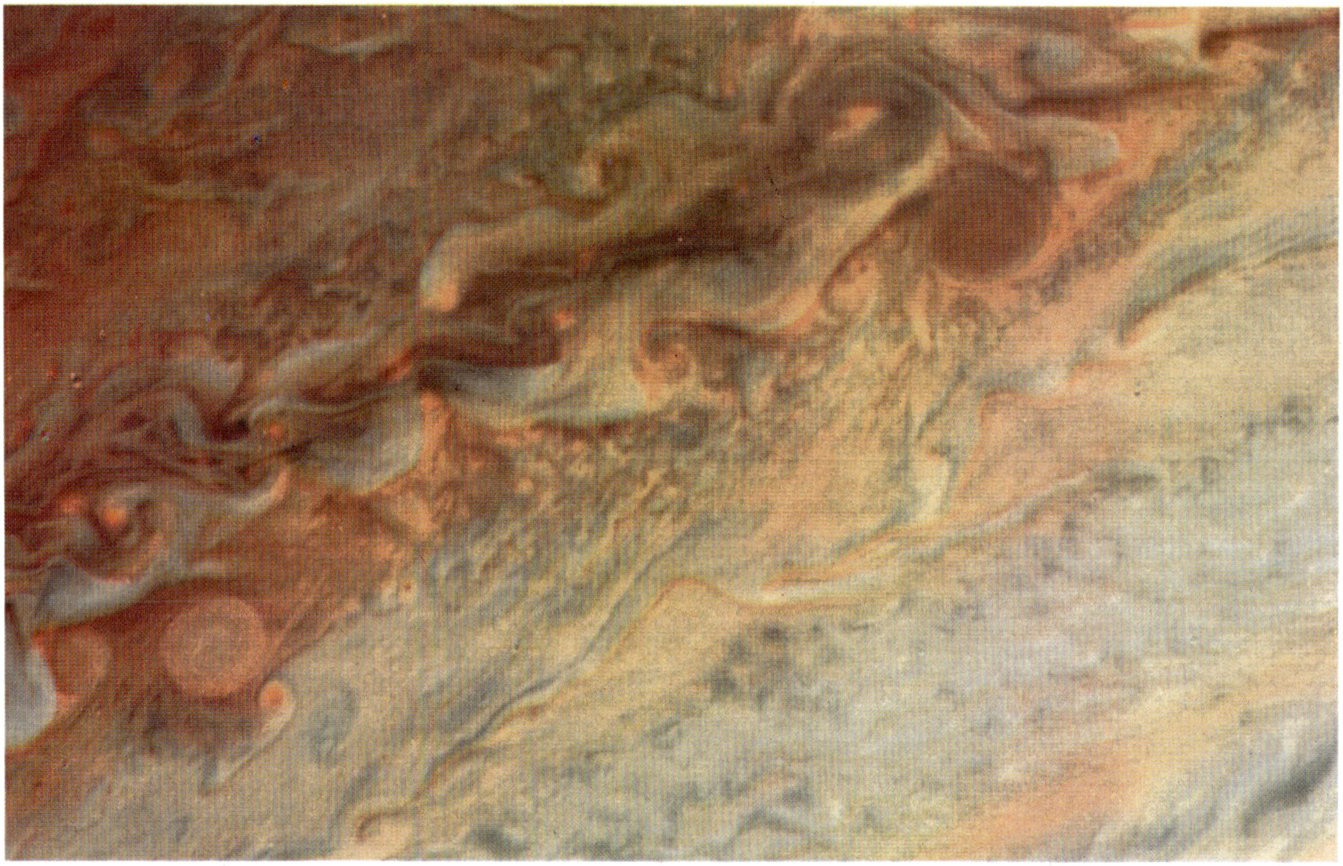

Great Red Spot
Voyager 2 (top)
3 July 1979

Great Red Spot
Voyager 1 (bottom)
4 May 1979

Cylindrical projections
Voyager 1 (top)
1 February 1979

Voyager 2 (bottom)
23 May 1979

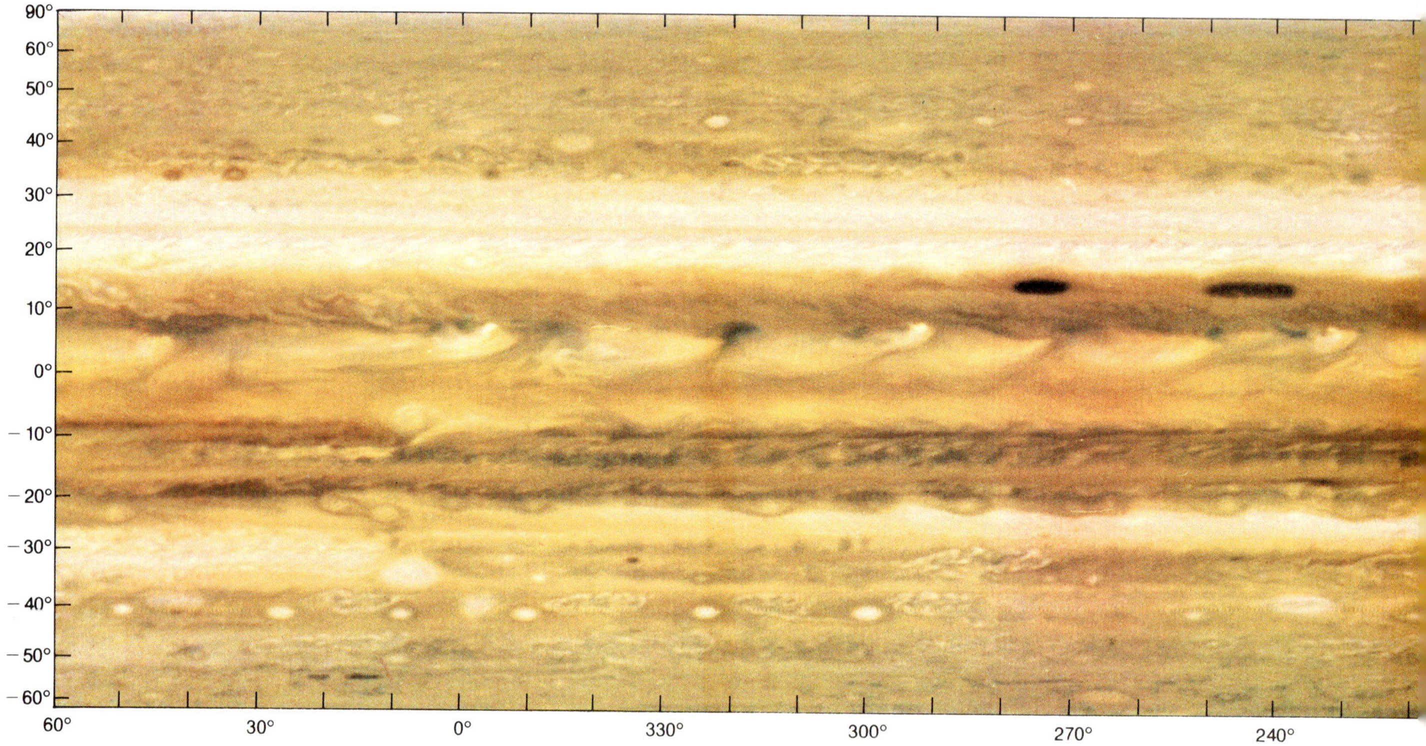

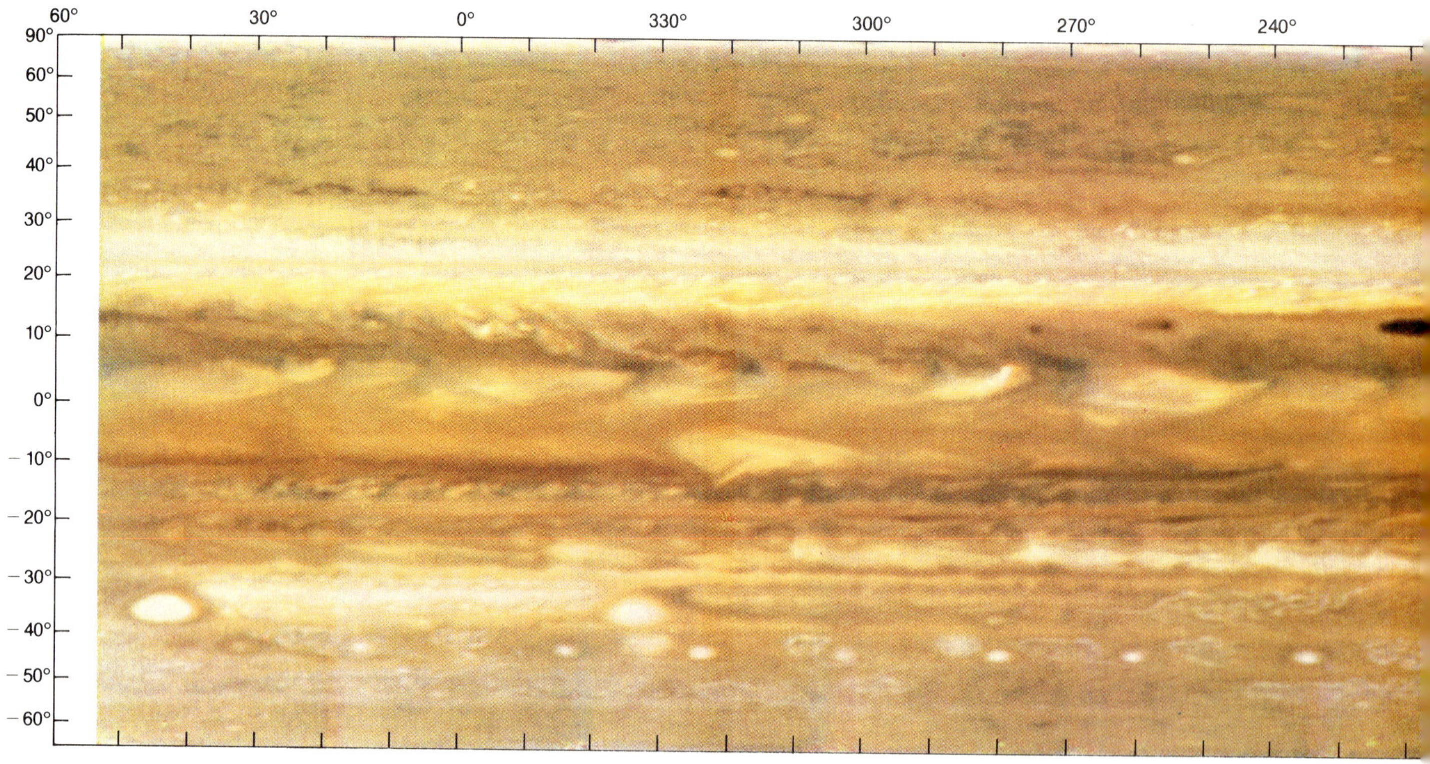

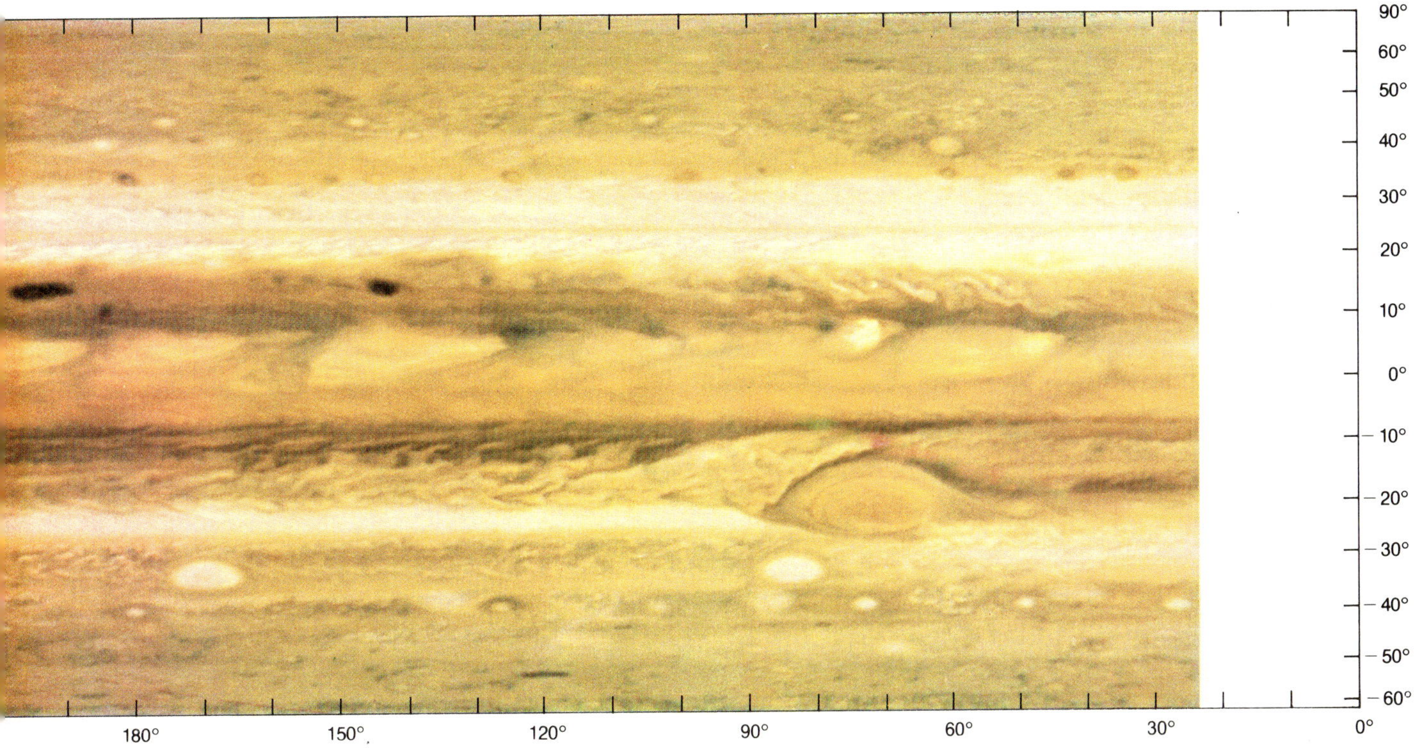
90°
60°
50°
40°
30°
20°
10°
0°
10°
20°
30°
40°
50°
60°
180°
150°
120°
90°
60°
30°
0°

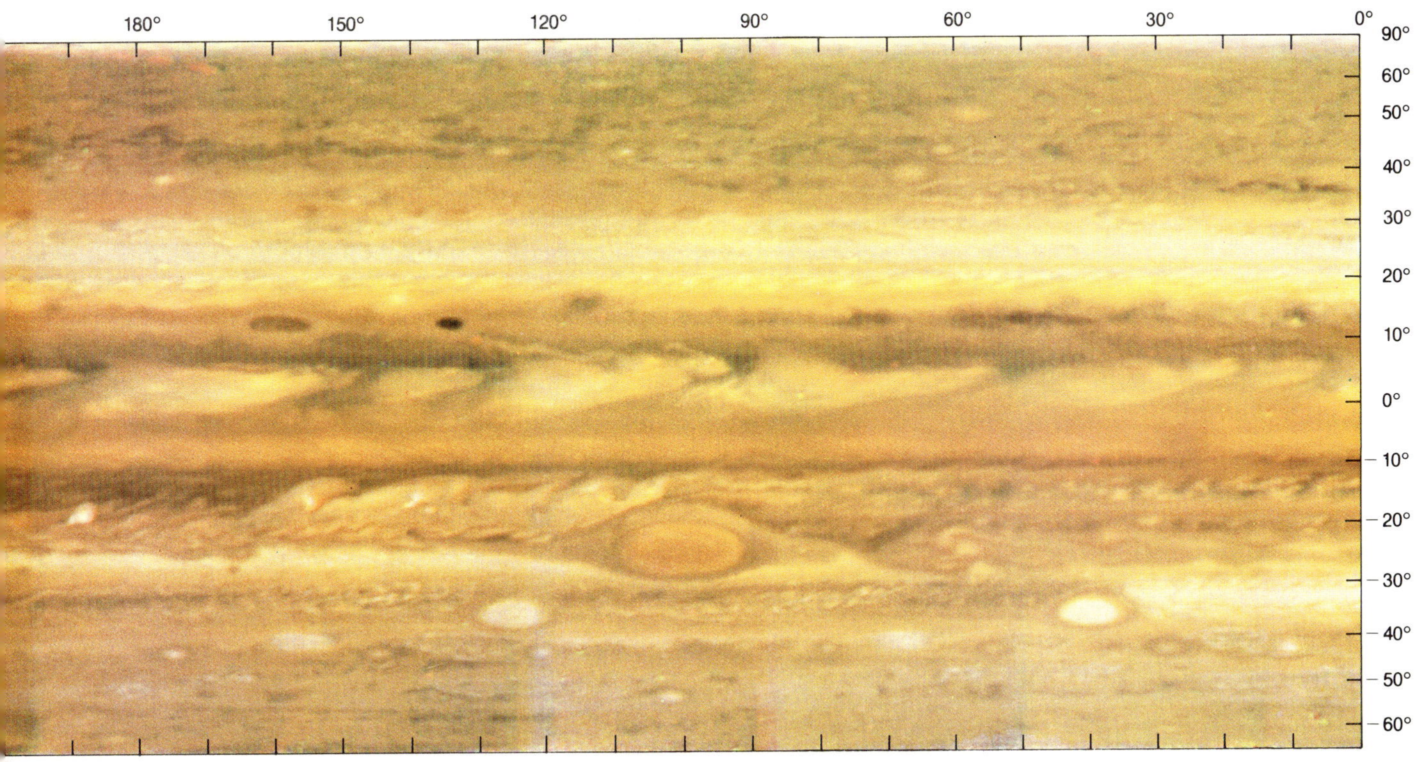
180°
150°
120°
90°
60°
30°
0°
90°
60°
50°
40°
30°
20°
10°
0°
10°
20°
30°
40°
50°
60°

North polar projection
Voyager 1
1 February 1979

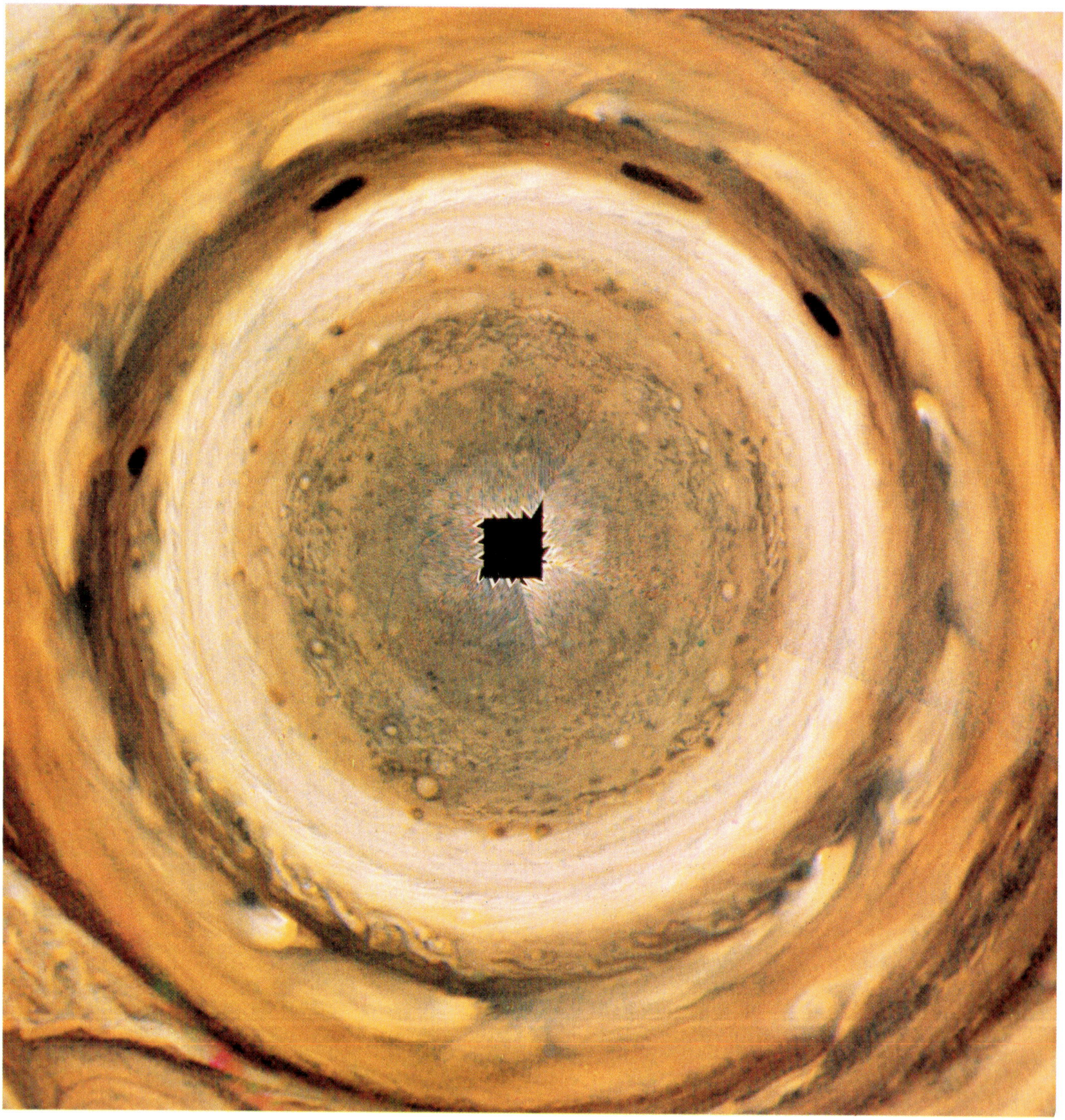

South polar projection
Voyager 1
1 February 1979

Io
Voyager 1
4 March 1979

Io
Voyager 1
4 March 1979

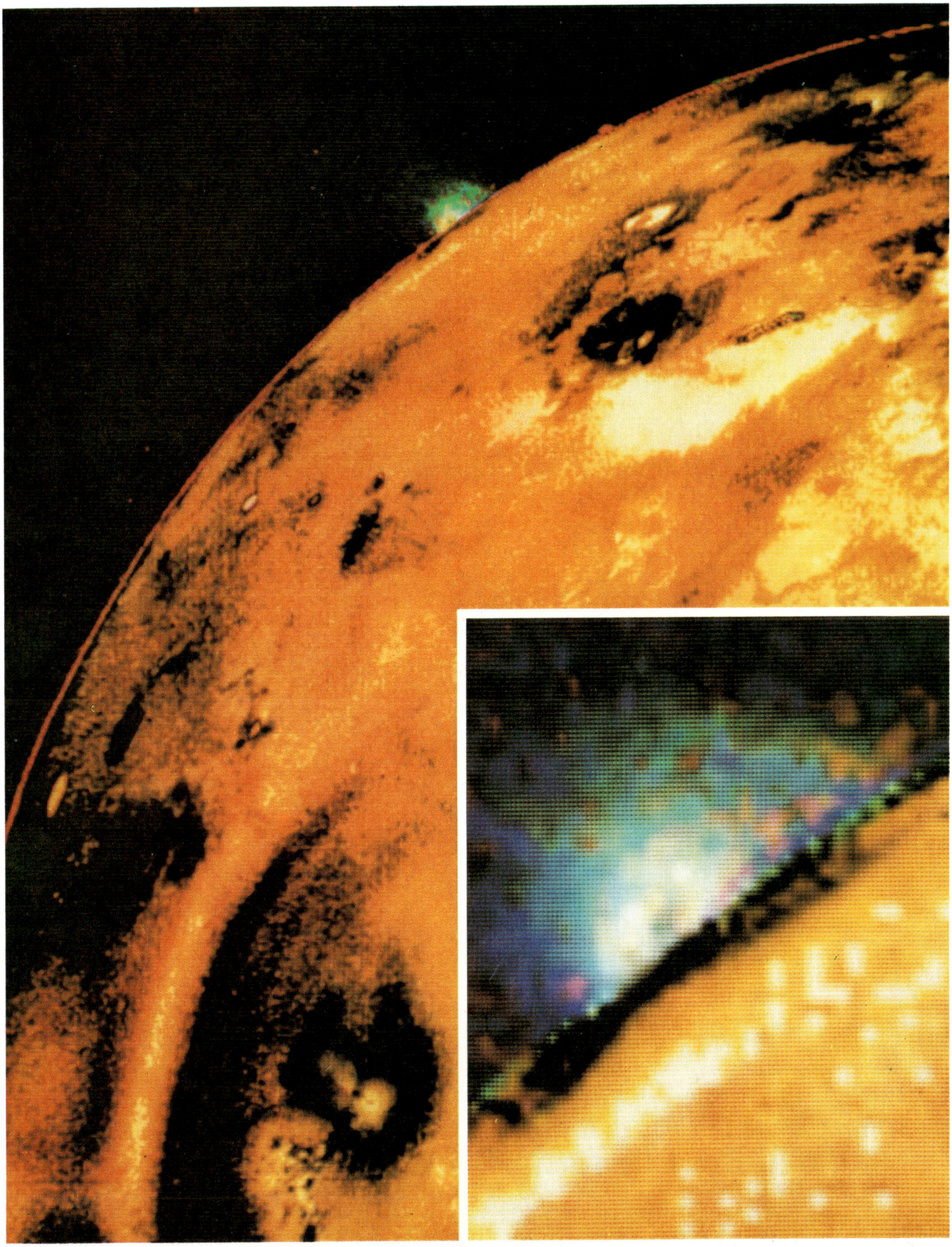

Europa
Voyager 2
8 July 1979

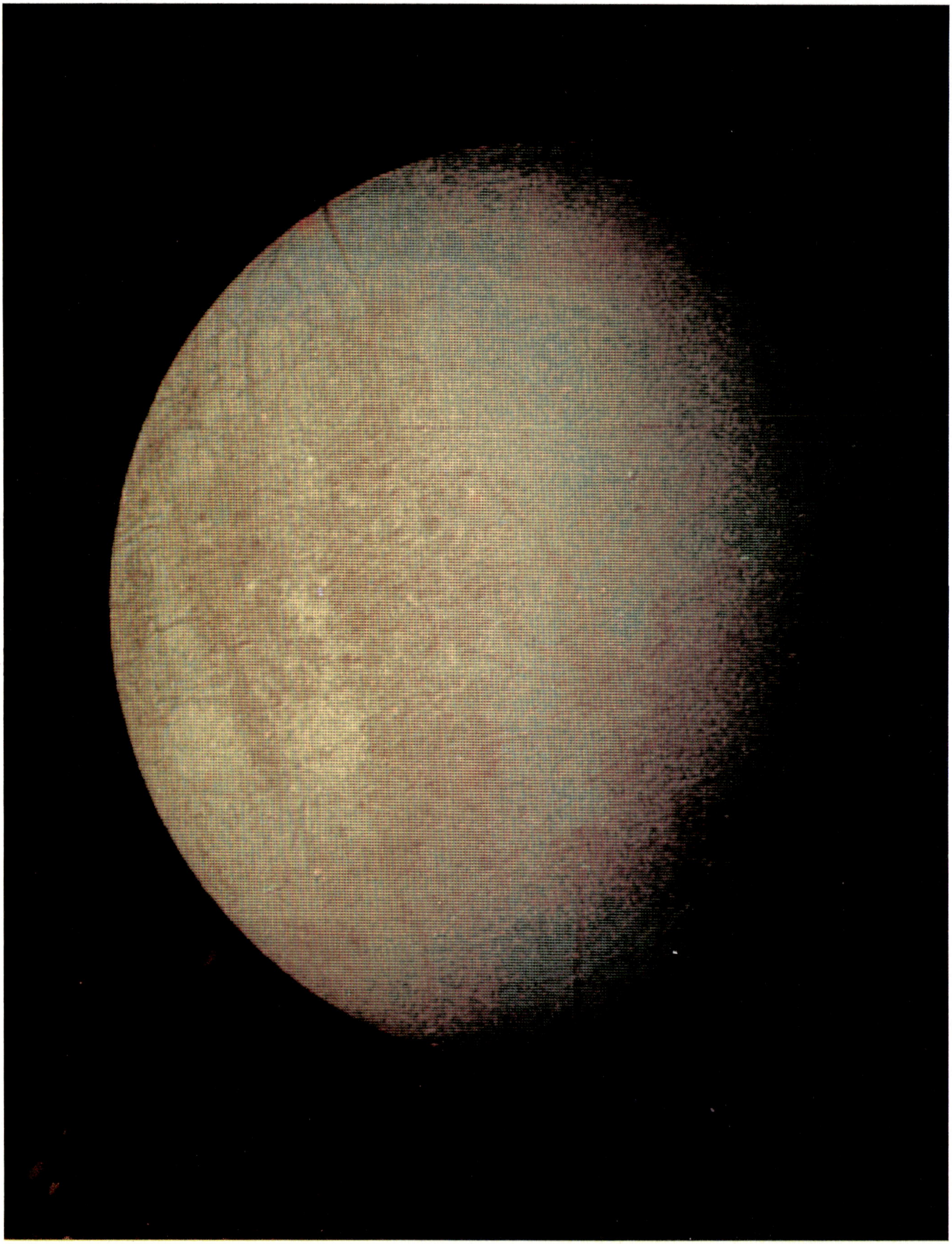

Europa
Voyager 2
9 July 1979

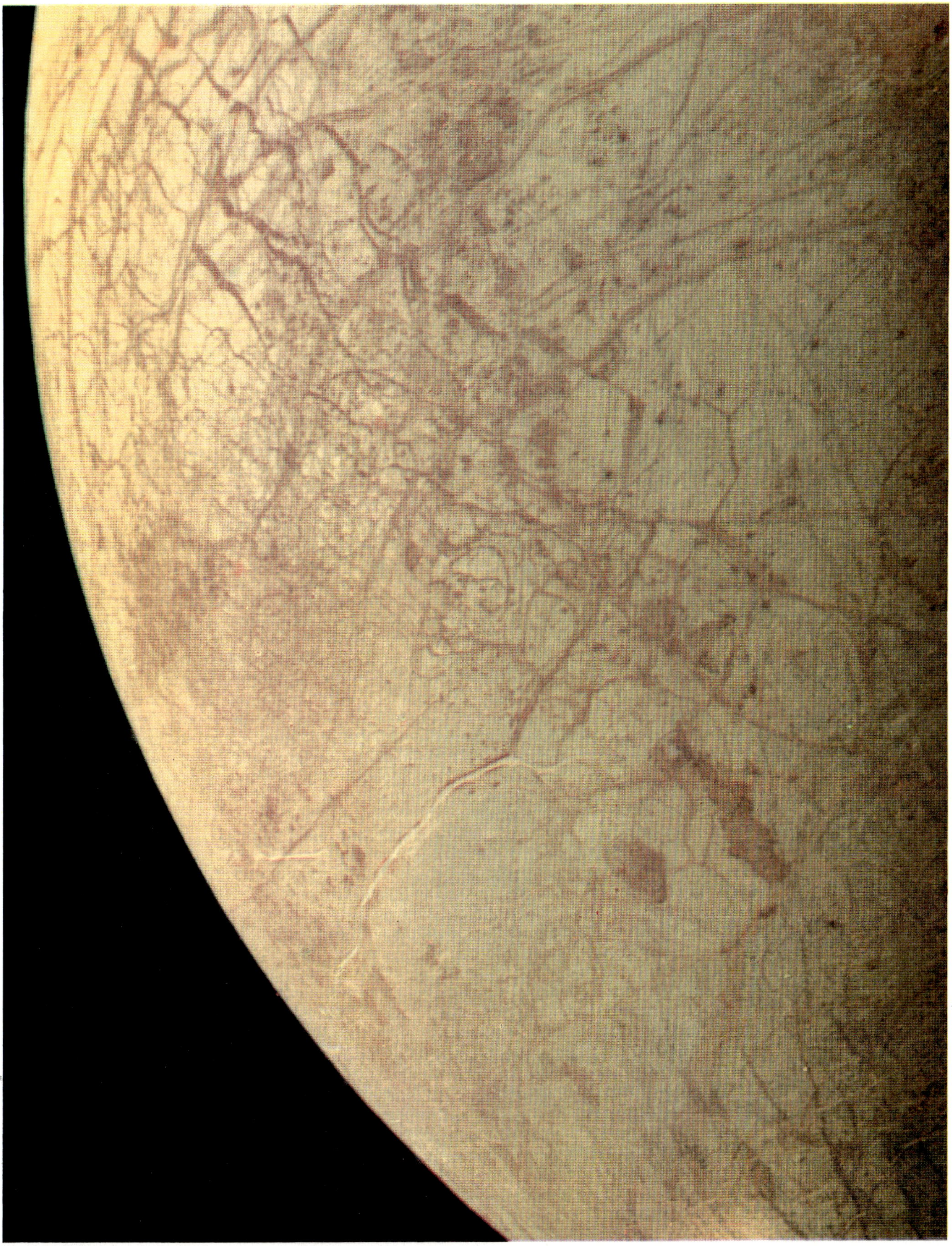

Ganymede
Voyager 2
7 July 1979

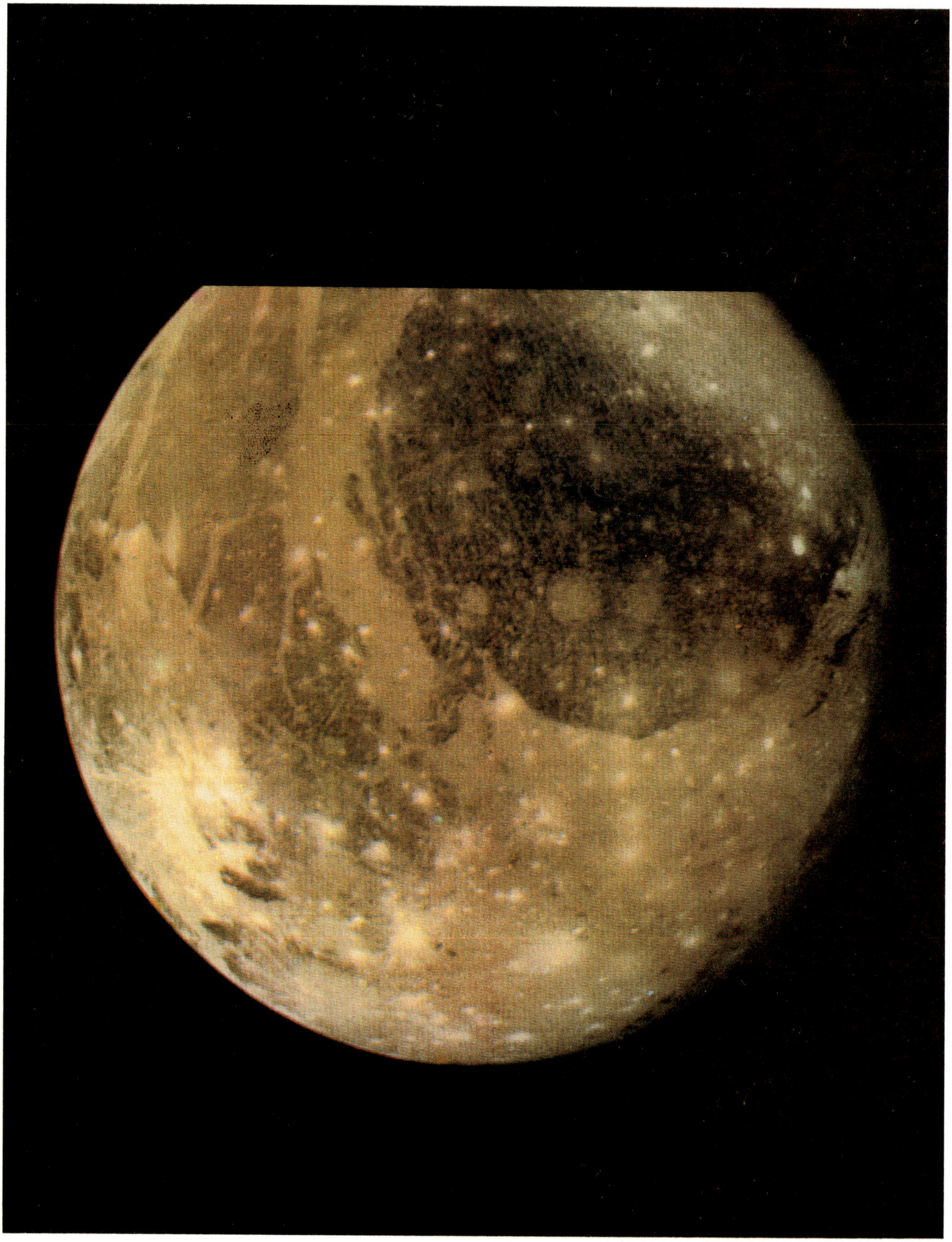

Ganymede
Voyager 2
8 July 1979

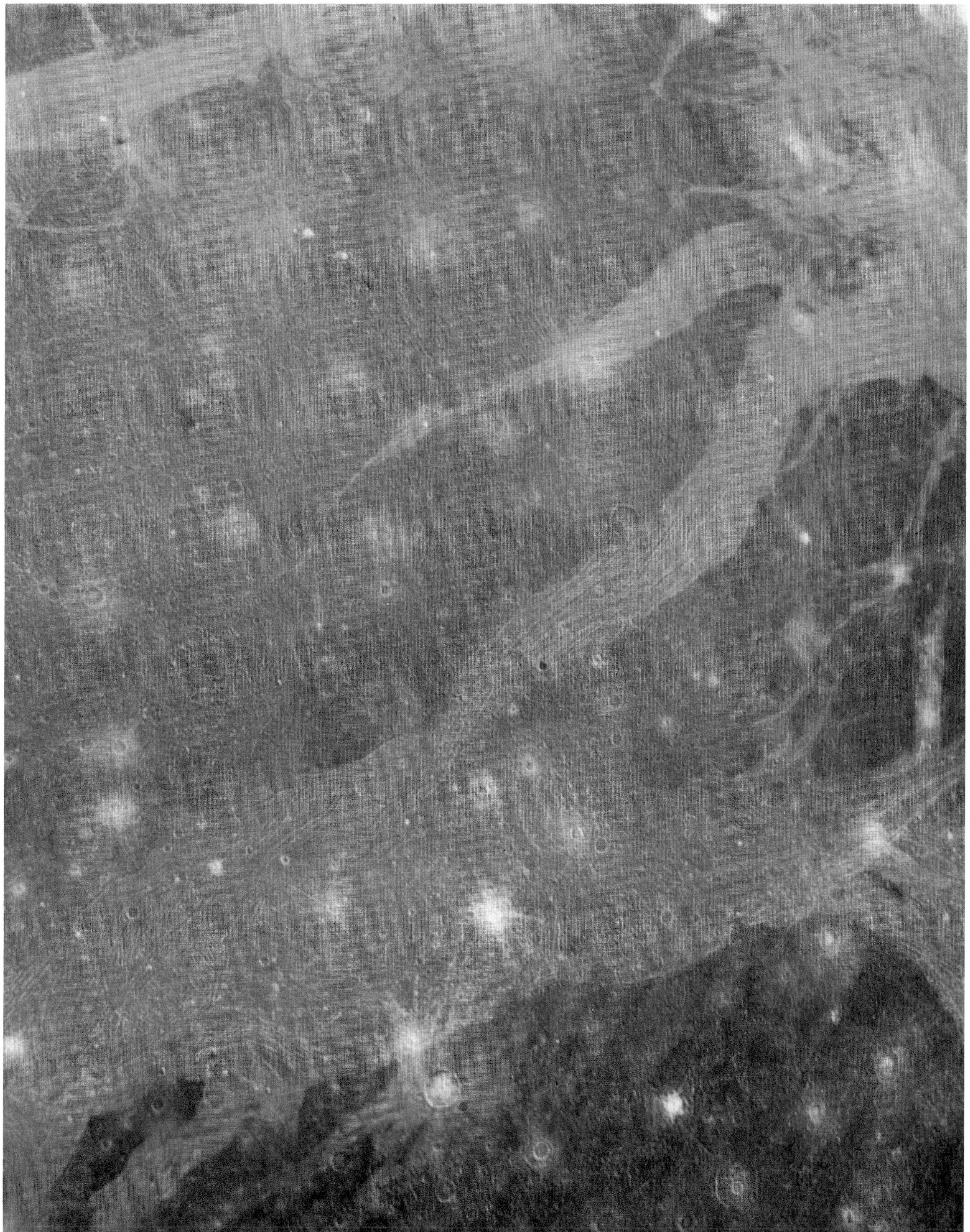

Callisto
Voyager 1
5 March 1979

Notes to Color Plates

Page 33. Three of the Galilean satellites can be seen in this image (Callisto is just visible in the lower left-hand corner). The spacecraft was at a distance of 28.4 million km from Jupiter.

Page 34. Earth-based image from Mt. Lemmon Observatory, using a 1.54 m telescope. University of Arizona.

Page 35. Jupiter's appearance was distinctly calmer at the time of the Pioneer flybys (top) than it appears in the Voyager images (bottom). Significant changes have taken place in the four months that separate these two Voyager images, particularly in the region of the GRS. The large white oval just below the GRS has drifted a considerable distance to the east (*see* pages 10–11).

Page 36 (Top). The pale orange line cutting the right-hand corner of the image marks the North Temperate Current, in which wind-speeds reach 120 m s^{-1}. A weaker jet further to the north (towards the top of the image) exhibits swirling cloud patterns. (Bottom) Colors have been exaggerated in this image to show up more detail. The NEB with a long-lived dark cloud feature (*see* page 29) cuts across the top of the image. Three of the plume-like features in the EZ can be seen immediately below the NEB.

Page 37 (Top) The white oval just below the GRS in this image is different from the one seen in a similar position at the time of the Voyager 1 encounter.
(Bottom) Blue and red have been deliberately exaggerated in this image to bring out details in the structure of the GRS.

Pages 38/39. These cylindrical projections of Jupiter clearly reveal the longitudinal drift of features on Jupiter. The two images are aligned in such a way that the scale of longitude is the same for both of them. The GRS has moved westward by about 30°, while other features have drifted in both directions, some of them at higher speeds.

Pages 40/41. This pair of polar stereographic projections shows up features situated at high latitudes. The black shapes at the poles themselves are due to lack of data.

Page 42. Io is shown from a distance of about 862,000 km. In the center of the image is the volcano Prometheus.

Page 43. Io at a distance of about 490,000 km. The inset shows a computer-enhanced image of the volcanic plume seen on Io's limb.

Page 44. Europa at a distance of 1.2 million km. The colors in this image have been slightly enhanced. The bluish areas in the polar regions should in fact appear white.

Page 45. Europa at a distance of 241,000 km.

Page 46. Ganymede at a distance of 1.2 million km. The large darkish region towards the north-east (top right) is Galileo Regio.

Page 47. Ganymede at a distance of 312,000 km. The smallest features visible in this image are about 5 km across.

Page 48. Callisto at a distance of 1.2 million km. The large ringed structure known as "Valhalla" can be seen near the limb in the upper left portion of the image.

The Satellites Introduction

Jupiter's satellite family, consisting of 16 known satellites, is the most extensive in the Solar System. The four largest satellites (Io, Europa, Ganymede and Callisto) form a group known as the "Galileans", so called because they were observed by Galileo Galilei in the very first days of telescopic research. All the Galileans have synchronous rotation (in other words, each satellite has a period of revolution around Jupiter equal to the period of the rotation of the satellite on its own axis). Four small satellites, Amalthea (discovered in 1892) and three others (1979 J1–Adrastea, 1979 J2 and 1979 J3) found on the Voyager images, move within the orbit of Io. Beyond Callisto, the outermost of the Galileans, there are eight more small satellites which may be asteroidal in nature. Their names, in order of increasing distance from Jupiter, are Leda, Himalia, Lysithea, Elara, Ananke, Carme, Pasiphaë and Sinope. Of these, the outermost four have retrograde motion. Neither the Pioneer nor the Voyager missions provided any information about the outer satellites.

In 1975 Charles Kowal, at Palomar, reported an extra asteroidal satellite of magnitude 21, but so far it has not been confirmed. No doubt other small satellites await discovery.

Apart from their intrinsic interest, Jupiter's satellites have been of scientific value in that their motions can be used to determine the mass of the planet and provide information about its gravitational field.

The Galileans data

	Io	Europa	Ganymede	Callisto
Mass (Jupiter = 1)	4.696×10^{-5}	2.565×10^{-5}	7.845×10^{-5}	5.603×10^{-5}
Mass (Moon = 1)	1.213	0.663	2.027	1.448
Mean density (kg m^{-3})	3,530	3,030	1,930	1,790
Mean surface gravity (m s^{-2})	1.80	1.46	1.43	1.14
Escape velocity (km s^{-1})	2.56	2.09	2.75	2.38

1. Satellite orbits
Jupiter's satellites fall broadly into three groups. The innermost group, comprising the Galileans together with Amalthea and the three newly discovered satellites, move in very nearly circular orbits in Jupiter's equatorial plane. Another four satellites, Leda, Himalia, Lysithea and Elara, all have a mean orbital distance in the range of 11 million km and make up the second group. Their orbits are more eccentric than those of the inner group and are inclined to them by an angle of slightly less than 30°. The remaining satellites, Ananke, Carme, Pasiphaë and Sinope, form the third group. They move in retrograde orbits, inclined to Jupiter's equatorial plane by an angle of approximately 150° to 160°. All the members of the outermost group orbit Jupiter at a mean distance of over 21 million km, thus the orbits are strongly perturbed by the Sun.

1

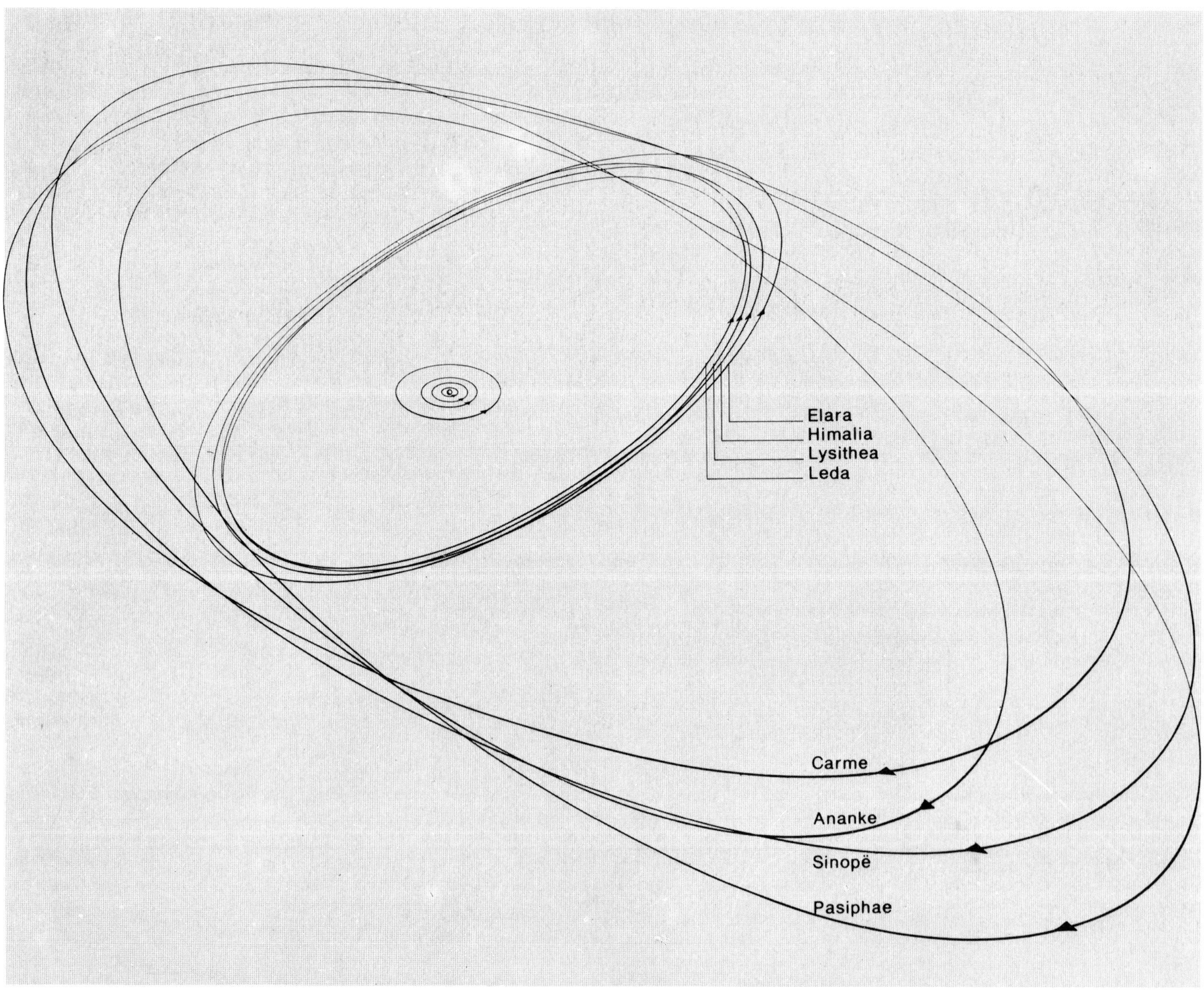

The Jovian Satellites

No.	Satellite	Discoverer	Year of discovery	Mean distance from Jupiter km	Diameter km	Magnitude	Orbital inclination degrees	Orbital eccentricity	Sidereal period days	Mean synodic period d hr min sec
XVI	1979 J3	Synnott	1980	127,600	~40	?	?	?	0.295	?
XIV	1979 J1	Jewitt & Danielson	1979	~128,400	~35	<20	?	?	0.297	?
V	Amalthea	Barnard	1892	181,300	155 × 270	14.1	0.4	0.003	0.498	0 11 57 27.6
XV	1979 J2	Synnott	1980	225,000	~75	<20	?	?	0.678	?
I	Io	Galileo, Marius	1610	421,600	3,632	4.9	0.0	0.0001	1.769	1 18 28 35.9
II	Europa	Galileo, Marius	1610	670,900	3,126	5.3	0.5	0.0001	3.551	3 13 17 53.7
III	Ganymede	Galileo, Marius	1610	1,070,000	5,276	4.6	0.2	0.0014	7.155	7 03 59 35.9
IV	Callisto	Galileo, Marius	1610	1,883,000	4,820	5.6	0.2	0.0074	16.689	16 18 05 06.9
XIII	Leda	Kowal	1974	11,100,000	8	20	26.7	0.1478	238.7	254
VI	Himalia	Perrine	1904	11,470,000	170	13.5	28	0.1580	250.6	266
X	Lysithea	Nicholson	1938	11,710,000	19	18.4	29	0.1074	259.2	276
VII	Elara	Perrine	1905	11,743,000	80	15.8	28	0.2072	259.7	276
XII	Ananke	Nicholson	1951	20,700,000	17	18.6	147	0.169	631	551
XI	Carme	Nicholson	1938	22,350,000	24	17.9	163	0.207	692	597
VIII	Pasiphaë	Melotte	1908	23,300,000	27	18.6	148	0.410	744	635
IX	Sinope	Nicholson	1914	23,700,000	21	18.1	157	0.275	758	645

2. Scale of the Galileans
The Galilean satellites are comparable in size with the smaller planets; Ganymede, for example, is slightly larger in diameter than Mercury. With the exception of Europa, all are larger than the Moon.

2

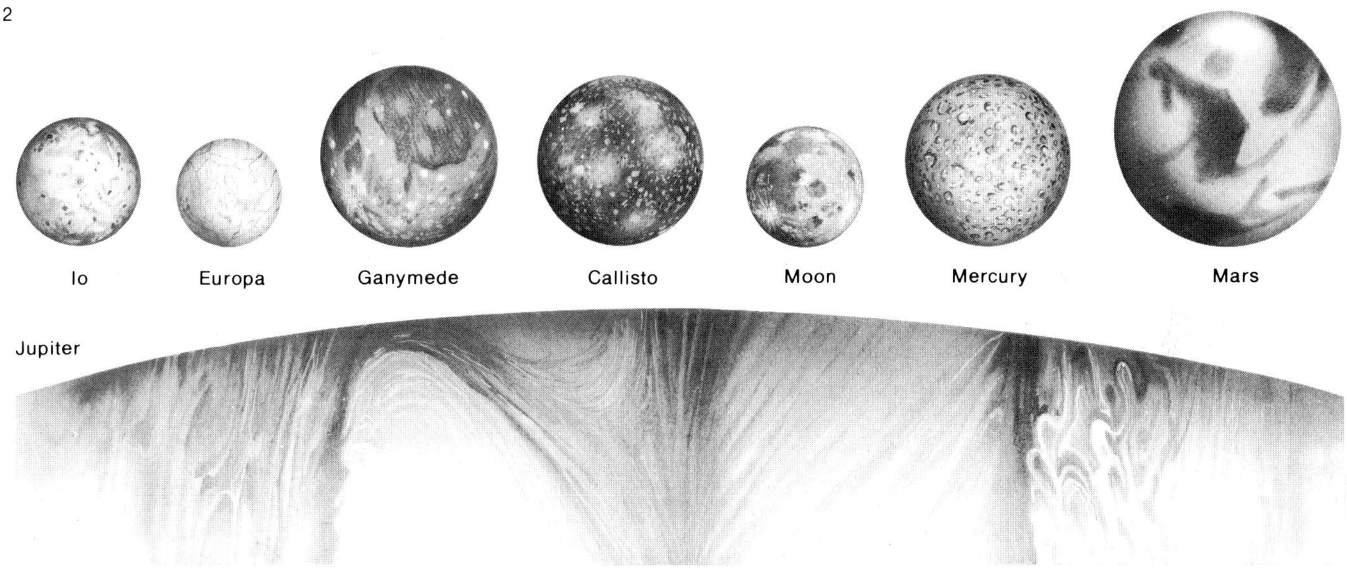

3

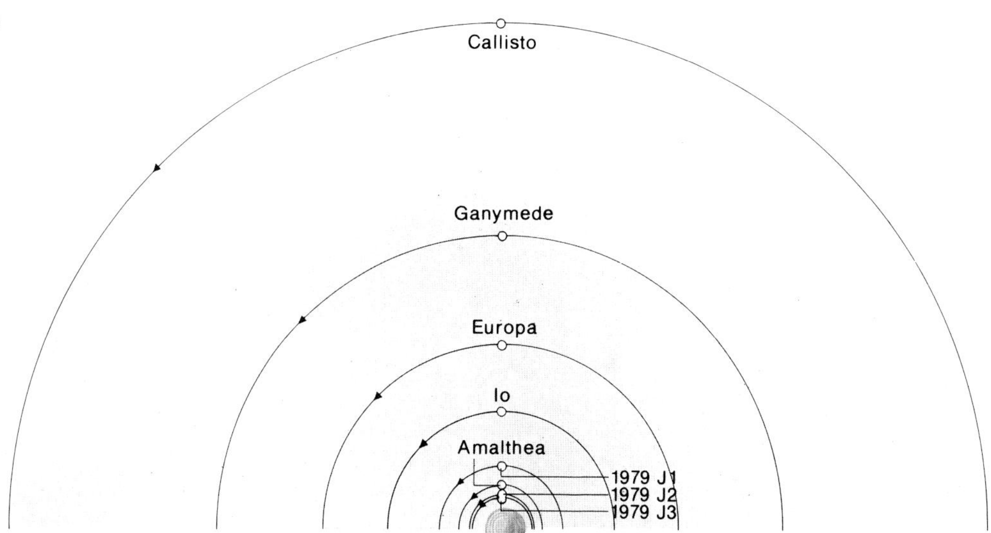

3. The Galileans
The orbits of the four largest and most important Jovian satellites lie within 2 million km of the planet. The satellites 1979 J1, 1979 J2, 1979 J3 and Amalthea are also shown, moving around Jupiter at distances of less than 400,000 km.

The Satellites History

Jupiter's four main satellites were observed by Galileo Galilei for the first time on 7 January 1610. It is possible that they had been seen even earlier by Simon Marius, and it was, in fact, Marius who gave the satellites their familiar names; but until recently these names were not used officially mainly because of the question of priority of discovery, and instead the satellites were referred to as I, II, III and IV. However, Galileo's work was much more reliable and extensive, and he is generally given the principal credit for discovering them. In *Sidereus Nuncius* (1610) he describes the occasion as follows:

"... I should disclose and publish to the world the occasion of discovering and observing four Planets, never seen from the beginning of the world up to our own times, their positions, and the observations made during the last two months about their movements and their changes of magnitude; and I summon all astronomers to apply themselves to examine and determine their periodic times, which it has not been permitted me to achieve up to this day ... On the 7th day of January in the present year, 1610, in the first hour of the following night, when I was viewing the constellations of the heavens through a telescope, the planet Jupiter presented itself to my view, and as I had prepared for myself a very excellent instrument, I noticed a circumstance which I had never been able to notice before, namely that three little stars, small but very bright, were near the planet; and although I believed them to belong to the number of the fixed stars, yet they made me somewhat wonder, because they seemed to be arranged exactly in a straight line, parallel to the ecliptic, and to be brighter than the rest of the stars, equal to them in magnitude ... When on January 8th, led by some fatality, I turned again to look at the same part of the heavens, I found a very different state of things, for there were three little stars all west of Jupiter, and nearer together than on the previous night."

Following further observations, Galileo wrote:

"I therefore concluded, and decided unhesitatingly, that there are three stars in the heavens moving about Jupiter, as Venus and Mercury round the Sun; which was at length established as clear as daylight by numerous other subsequent observations. These observations also established that there are not only three, but four, erratic sidereal bodies performing their revolutions around Jupiter."

There are reasons for claiming that this was the most important of all Galileo's discoveries. It showed that contrary to the old beliefs there were at least two centers of movement in the universe; in other words, not everything revolved around the Earth. Some church officials were decidedly unenthusiastic, and one opponent refused to look at the satellites through Galileo's telescope—which he believed to be bewitched. Galileo is said to have expressed the hope that the official concerned would have a good view of the Jovian system on his way to Heaven.

Theoretical questions

Before the advent of the first practical marine chronometer in the eighteenth century, the phenomena of the Galileans (*see* pages 54–55) were regarded as potentially important for the determination of longitude by marine navigators far from land, since the times of the phenomena could be predicted accurately. The method, however, was never actually used with any real success.

More importantly, it was by comparing the times of eclipses of the Galileans that the Danish astronomer Ole Römer established that the velocity of light was finite, and made the first reasonably accurate measurement of it. He realized that the discrepancies between predicted and observed times of satellite phenomena were due to Jupiter's varying distance from Earth, and obtained a value for the velocity of light accurate to within 2 percent.

1. Galileo Galilei (1564–1642)
Among Galileo's most significant contributions to science was the discovery of the four Jovian satellites that are now named in his honor. They were originally called by him "Medicean planets", after the Medici family.

2. Sidereus Nuncius
Galileo published the first results of his telescopic researches in 1610, in *Sidereus Nuncius*, two pages of which are shown here. Jupiter is represented by a large star, surrounded by the Galileans, illustrated as smaller black stars.

2

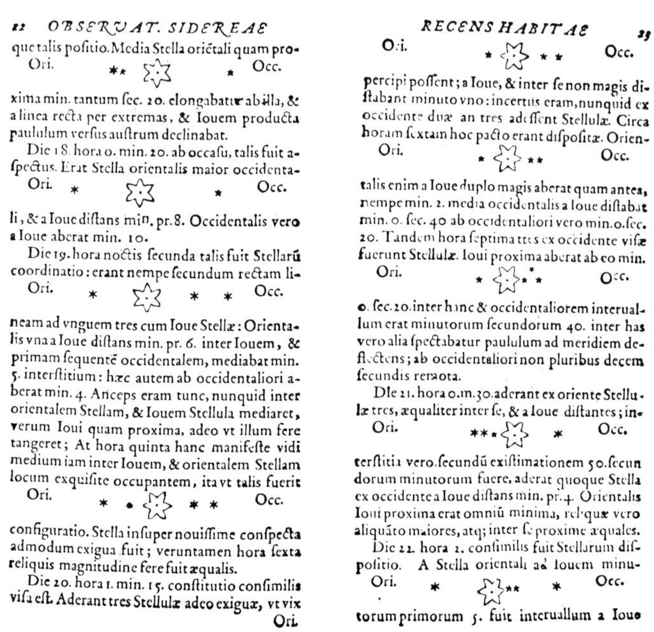

22 OBSERVAT. SIDEREAE

que talis poſitio. Media Stella oriẽtali quam pro-

Ori. Occ.

xima min. tantum ſec. 20. elongabatur ab illa, & a linea recta per extremas, & Iouem producta paululum verſus auſtrum declinabat.

Die 18. hora 0. min. 20. ab occaſu, talis fuit aſpectus. Erat Stella orientalis maior occidenta-

Ori. Occ.

li, & a Ioue diſtans min. pr. 8. Occidentalis vero a Ioue aberat min. 10.

Die 19. hora noctis ſecunda talis fuit Stellarũ coordinatio: erant nempe ſecundum rectam li-

Ori. Occ.

neam ad vnguem tres cum Ioue Stellæ: Orientalis vna a Ioue diſtans min. pr. 6. inter Iouem, & primam ſequentẽ occidentalem, mediabat min. 5. interſtitium: hæc autem ab occidentaliori aberat min. 4. Anceps eram tunc, nunquid inter orientalem Stellam, & Iouem Stellula mediaret, verum Ioui quam proxima, adeo vt illum fere tangeret; At hora quinta hanc manifeſte vidi medium iam inter Iouem, & orientalem Stellam locum exquiſite occupantem, ita vt talis fuerit

Ori. Occ.

configuratio. Stella inſuper nouiſſime conſpecta admodum exigua fuit; veruntamen hora ſexta reliquis magnitudine fere fuit æqualis.

Die 20. hora 1. min. 15. conſtitutio conſimilis viſa eſt. Aderant tres Stellulæ adeo exiguæ, vt vix

Ori.

RECENS HABITAE 23

Ori. Occ.

percipi poſſent; a Ioue, & inter ſe non magis diſtabant minuto vno: incertus eram, nunquid ex occidente duæ an tres adeſſent Stellulæ. Circa horam ſextam hoc pacto erant diſpoſitæ. Orien-

Ori. Occ.

talis enim a Ioue duplo magis aberat quam antea, nempe min. 2. media occidentalis a Ioue diſtabat min. 0. ſec. 40 ab occidentaliori vero min. 0. ſec. 20. Tandem hora ſeptima tres ex occidente viſæ fuerunt Stellulæ. Ioui proxima aberat ab eo min.

Ori. Occ.

0. ſec. 20. inter hanc & occidentaliorem interuallum erat minutorum ſecundorum 40. inter has vero alia ſpectabatur paululum ad meridiem deflectens; ab occidentaliori non pluribus decem ſecundis remota.

Die 21. hora 0. m. 30. aderant ex oriente Stellulæ tres, æqualiter inter ſe, & a Ioue diſtantes; in-

Ori. Occ.

terſtitia vero ſecundũ exiſtimationem 50. ſecundorum minutorum fuere, aderat quoque Stella ex occidente a Ioue diſtans min. pr. 4. Orientalis Ioui proxima erat omniũ minima, [illegible] vero aliquãto maiores, atq; inter ſe proxime æquales.

Die 22. hora 2. conſimilis fuit Stellarum diſpoſitio. A Stella orientali ad Iouem minu-

Ori. Occ.

torum primorum 5. fuit interuallum a Ioue

3

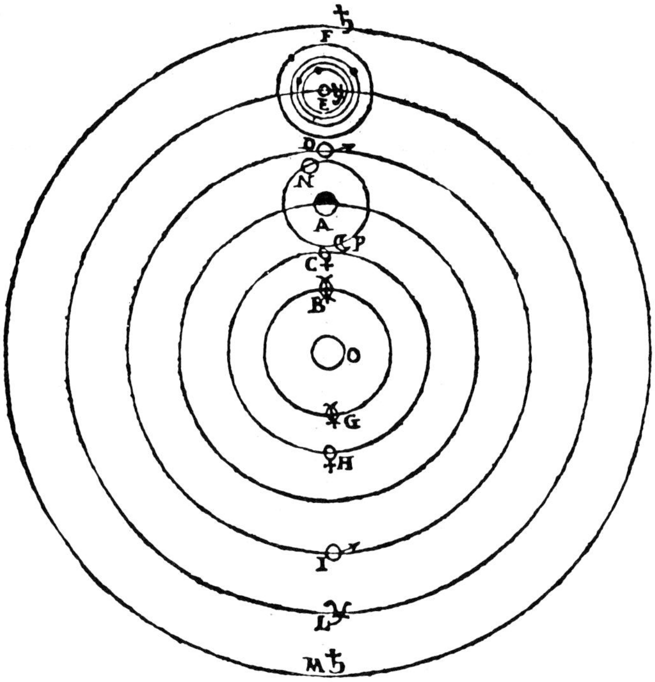

Maps of the Galileans

In 1961 A. Dollfus and his colleagues at the Pic du Midi Observatory in France produced some preliminary mercator maps of Io, Europa, Ganymede and Callisto. Although inevitably very rough, these maps at least demonstrate that a certain amount of surface detail can be seen from Earth using very large telescopes. The orange color of Io was also discernible. Since the Pioneer and Voyager missions, however, it has been possible to produce far more detailed and accurate maps based on the photographs sent back by the space-probes (*see* pages 58–59, 66–67, 70–71 and 78–79). There are still some gaps in the coverage, and not all the areas were photographed to the same resolution, so in some places the maps are more precise than in others.

4A

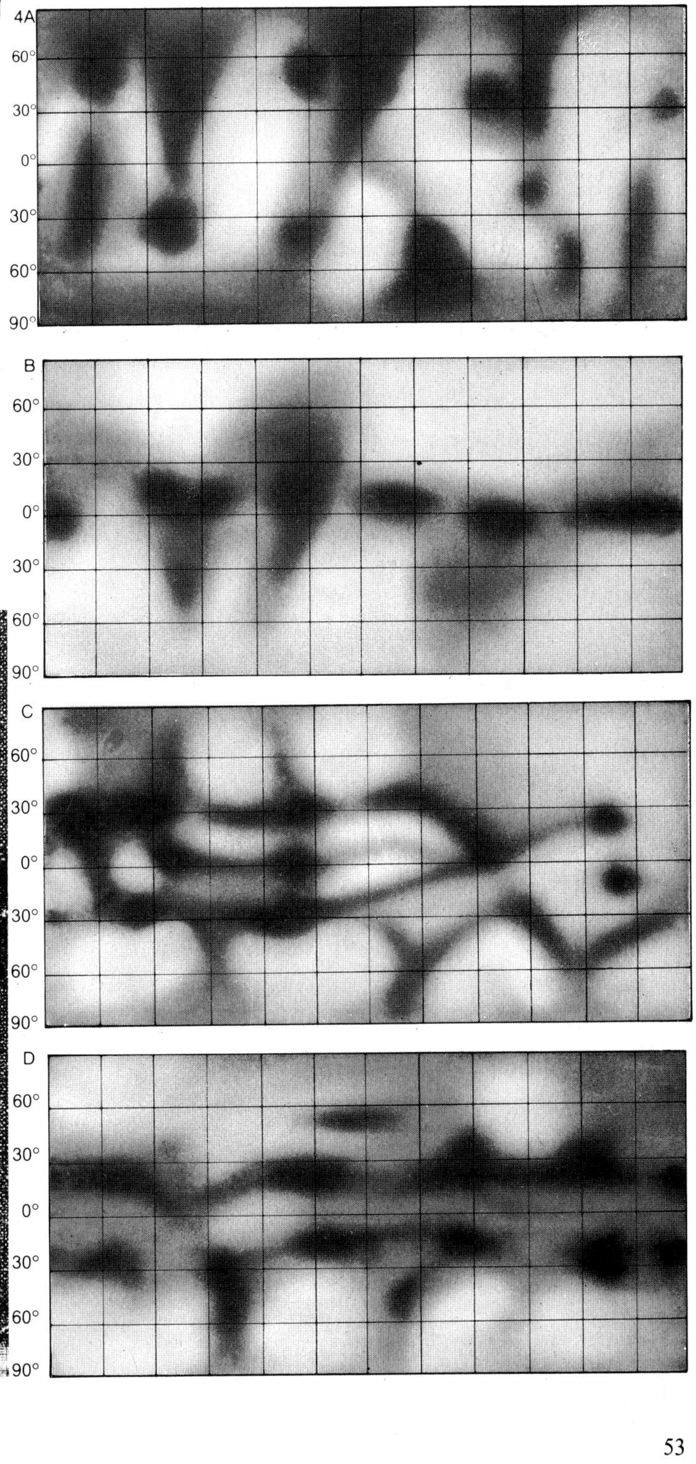

3. The Copernican system
Galileo's drawing of the Copernican system of the universe, published in *Dialogo* (1632), includes the Jovian satellites that he had discovered. Copernicus believed that the planets moved in circular orbits, with the Sun at the center.

4. Maps by Dollfus
Maps of Io (**A**), Europa (**B**), Ganymede (**C**) and Callisto (**D**) were made in 1961.

5. Ole Römer (1644–1710)
Römer, inventor of the transit instrument shown here, determined the velocity of light from observations of the times of eclipse of each Galilean.

5

The Satellites Observations

Any small telescope will show all four Galileans as definite discs. Amateur observers can take great pleasure in watching the satellites' phenomena: eclipses, occultations, transits and shadow transits. Because the orbits all lie practically in the plane of Jupiter's equator, which is itself inclined by only about 3°, the Galileans alternately transit across Jupiter's disc and are occulted behind it during the course of each revolution. Only Callisto is far enough away from Jupiter to pass clear of the disc, which it does when the declination of the Earth as seen from Jupiter is greater than about 2°40′. Generally, several satellites can be seen at any particular moment; it is rare for all four to be out of view simultaneously.

Shadow transits are relatively easy to observe, the shadows appearing as dark spots. As far as actual transits are concerned, the satellites behave differently. Near the start of a transit each satellite will appear as a bright spot, but may soon be lost to view until it reappears as a bright spot once again just before the transit ends. Io may, however, be seen as a dusky spot when a transit is well advanced, particularly when it is in front of a bright zone, while Europa may remain bright for most of the time, although it is clearly visible only when projected against a dark belt; Ganymede and Callisto are visible as grey spots during transit. It is interesting to compare the apparent size of a satellite in transit with its shadow. The shadow will seem the larger due to the effects of penumbra.

Mutual phenomena may also be observed. A partial occultation of Io by Callisto, for example, was seen on 18 February 1932 by two famous observers of Jupiter, W. H. Steavenson and B. M. Peek, while on 8 February 1920, C. S. Saxton, using only a 3 in (7.6 cm) refractor, saw an eclipse of Io by the shadow of Ganymede. Other instances are on record.

Observations by Pioneer and Voyager

The first attempt to study the Galileans from space-probes was made in December 1973, when Pioneer 10 encountered Jupiter. In December of the following year, images were sent back to Earth by Pioneer 11, but much more detailed information about the satellites was obtained from the two Voyager missions in 1979. The Voyager trajectories were designed to provide the fullest pictorial coverage of the satellite surfaces by taking advantage of their synchronous rotation and rapid motion around Jupiter. Thus Voyager I took pictures of the Jupiter-facing hemispheres of Ganymede and Callisto, passing them at high latitudes to achieve good coverage of their north poles, while Voyager 2 recorded the outward-facing hemispheres and passed further to the south of Ganymede than had Voyager 1; Voyager 2 also obtained images of part of Europa's surface. Both spacecraft were able to achieve detailed coverage of much of the surface of Io, thanks to its short orbital period.

1979 J3

Three inner satellites have been discovered by examination of the Voyager pictures of 1979. The most recently discovered is 1979 J3, which orbits Jupiter at a distance of 56,200 km above the cloud tops; it is thus the closest to Jupiter of the known satellites. It has a diameter of about 40 km.

1979 J1

Orbiting inside the path of Amalthea, 1979 J1, like 1979 J3, lies within the elusive Jovian ring, so that the two satellites may play an important role in the stability of the ring itself. 1979 J1 is apparently egg-shaped, with a diameter of between 30 and 40 km and an albedo of less than 0.05; it may therefore be a high-density object. It has a period of 7 hr 8 min, and moves at 57,000 km above the visible surface of Jupiter at the cloud tops.

Amalthea

Amalthea was discovered in 1892 by Edward Emerson Barnard, using the 36 in (91 cm) refractor at the Lick Observatory, while he

1

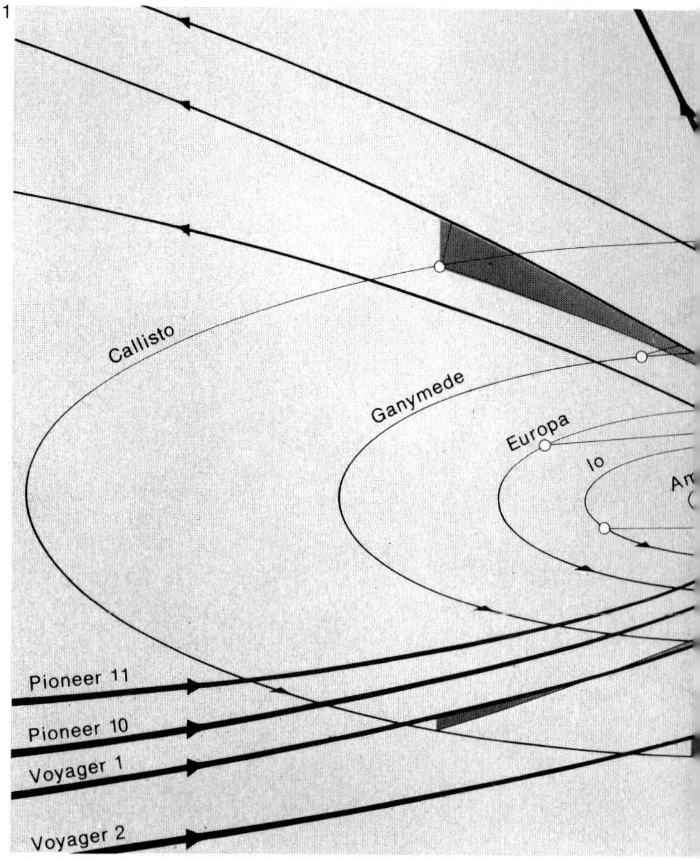

1. Spacecraft trajectories
The paths taken by each of the four spacecraft that have visited Jupiter were carefully designed to make optimum use of the time available for observation. Pioneers 10 and 11 concentrated mainly on Jupiter itself, sending back only a few low-definition images of the satellites. The Voyager vehicles, on the other hand, between them provided coverage of between about 50 and 90 percent of the surfaces of the Galileans at high resolution. Europa received the poorest coverage, Io the most extensive. Images of Amalthea were also obtained.

2. The Galileans
A small telescope or a good pair of binoculars is sufficient to reveal the Galilean satellites. These two Earth-based photographs were taken at McDonald Observatory on 25 and 30 January 1955.

2A

B

was carrying out a deliberate search for new satellites. Amalthea was in fact the very last satellite to be discovered by direct visual observation. It is so faint and so close to Jupiter that it is difficult to observe from Earth, but it has been carefully studied, and its orbit was well known even before the Pioneer and Voyager missions. Voyager I approached it to a distance of 420,100 km, and Voyager 2 to 558,270 km. Both probes took pictures of it.

Amalthea's orbit is close to the "Roche limit", the limiting distance within which an orbiting body with little gravitational cohesion would be disrupted by the pull of gravity. Amalthea is in fact elongated by the gravitational pull of Jupiter, and has a generally ellipsoidal shape. Its 270 km long axis is pointed towards Jupiter, while its 155 km short axis lies at right angles to the orbital plane. It has an irregular profile, which indicates that it must be fairly dense: a more plastic body would become smooth.

Amalthea has been found to be rather warmer than it would be if it were simply absorbing and re-radiating solar radiation and the radiation it receives from Jupiter. Possibly the additional heating is derived from electrical currents induced by the Jovian magnetic field (*see* pages 18–21). It is red in color, and the surface is probably covered with sulphur, which could only originate from Io.

Four features were shown on the Voyager images; two craters, known as Pan and Gaea, and two mountains known as Mons Ida and Mons Lyctas. Nothing definite is known about them.

As seen from Jupiter, Amalthea would have an apparent diameter of 7′24″. As the apparent diameter of the Sun as seen from Jupiter is less than 6′, Amalthea would be able to produce a total solar eclipse.

1979 J2

This was the second new satellite to be found on the Voyager pictures; it was discovered while the photographs were being examined to confirm 1979 J1. The new body is between 70 and 80 km in diameter, and orbits at 151,000 km above the cloud tops, so that it moves outside the orbit of Amalthea but within that of Io. Its revolution period is 16 hr 16 min, but at the moment nothing further is known about it.

3. Satellite phenomena
The various possible configurations of the satellites provide a focus of interest for observers. More common examples include shadow transits (**A**), occultations (**B**) and eclipses (**C**). Rarer events, such as mutual occultations and eclipses, may also be observed.

3A

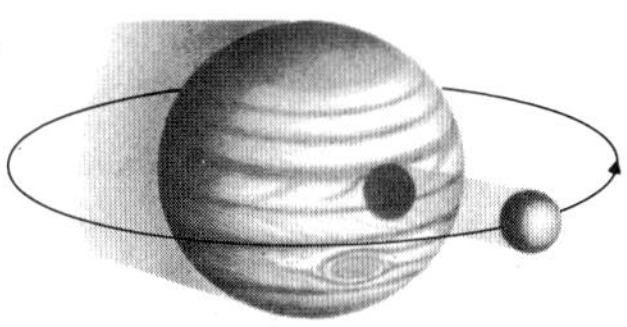

B

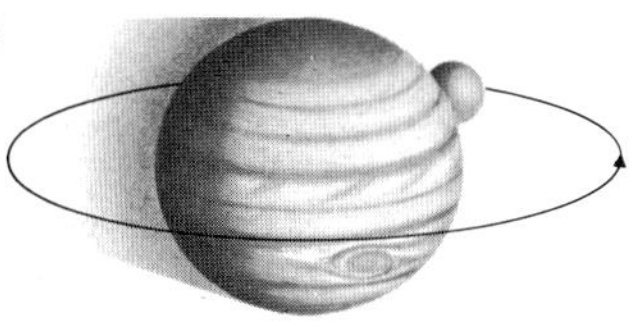

C

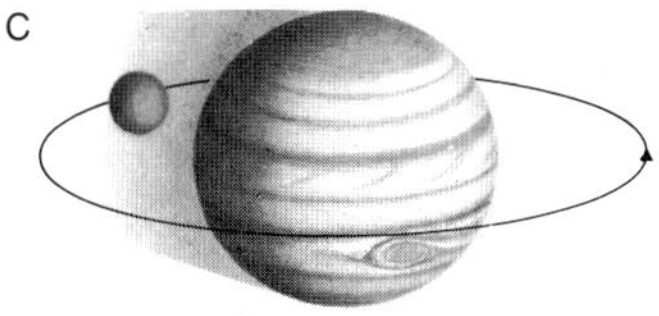

4

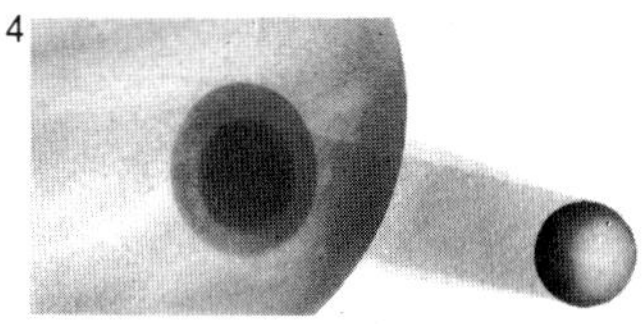

5

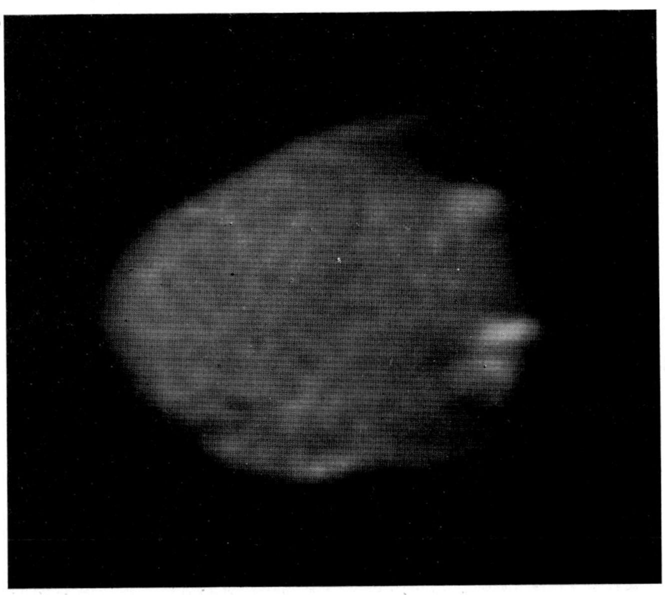

4. Effect of the penumbra
The shadow cast by a satellite appears larger than the satellite itself because of the area of partial shadow or "penumbra"

5. Amalthea
The irregular-shaped satellite Amalthea was photographed by Voyager 1 on 4 March 1979 from a distance of 425,000 km. Some of the indentations in the photograph may be craters.

6. Satellite 1979 J1
Jupiter's fourteenth satellite was discovered in this computer-enhanced image made by Voyager 2 on 8 July 1979. The satellite appears as a bright streak at the lower right of the image; the other streak is the track of a star. The grey band is Jupiter's ring.

6

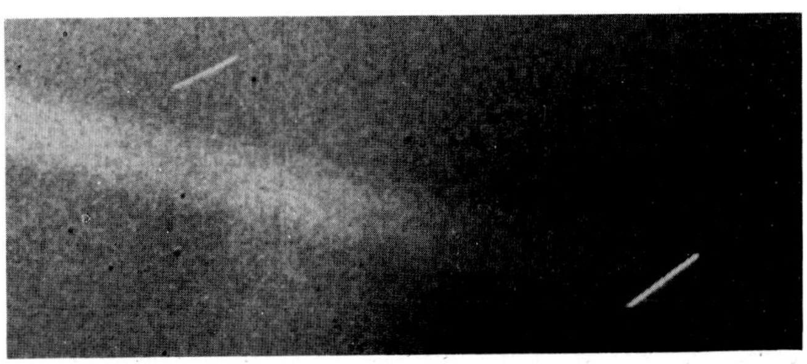

The Rings of Jupiter

The discovery that Jupiter is surrounded by rings of particles similar to those of Saturn and Uranus was another unexpected result of the Voyager mission. Jupiter's rings were first detected in a narrow-angle frame targeted halfway between Amalthea and the limb of Jupiter at 16 hr 52 min before the closest approach, as Voyager 1 crossed the equatorial plane of the Jovian system. Following this discovery, Voyager 2 was programmed to take additional pictures giving considerably better resolution.

The rings seem to consist of several components. The brightest, rather narrow portion has a radius at the outer edge of 126,380 $\pm$ 140 km (1.772 $\pm$ 0.002 R_J). There is also a narrow bright segment 800 $\pm$ 100 km wide with the inner edge at 125,580 $\pm$ 140 km (1.688 $\pm$ 0.002 R_J). The ring particles seem to extend nearly all the way down to the planet itself. This has other important consequences: the upper atmosphere would distribute the tiny particles on a global scale into a haze layer, which would then absorb the incoming solar radiation at visible and ultraviolet wavelengths. It is even possible that the ring material is a source of oxygen to the upper atmosphere that could be related to the carbon monoxide unexpectedly detected in Jupitêr's atmosphere.

The ring has a characteristically orange color and seems to be composed of particles of radius about 4 μm. This is in complete contrast with the Saturnian ring particles, which are estimated to be several centimeters in cross-section.

Origin and composition

The Jovian rings lie well inside the classical Roche limit for the break-up of a liquid body, which occurs at 2.44 R_J. It would therefore seem that the particles must be relatively high-density rocks (dust), rather than the icy material of the Saturnian rings. If, for example, the material is assumed to have roughly the same density as Io (3.5×10^3 kg m^{-3}), then the edge of the ring would be situated at 1.81 R_J, which is close to the observed position.

It is possible, therefore, that the material that currently resides in the Jovian ring originated from the gravitational break-up of a tiny inner satellite that evolved in the neighborhood of the Roche limit. This, however, would not create much material. Additional sources might include material from Io and other inner satellites, and debris from comets and meteorites.

The newly discovered inner satellites, 1979 J1 and 1979 J3, play an important role in relationship with the outer edge of the rings. 1979 J1 is situated precisely at the outer edge of the ring system and, as it moves around Jupiter, it sweeps up magnetospheric particles that reside there, enabling a sharp outer edge to the ring system to form. There is also the possibility that other small satellites of this type may reside in this region, and their resonant interactions may also account for the inner edge of the 5,200 km wide inner shelf. The discovery of 1979 J3 was not, therefore, a complete surprise.

A similar situation may account for the narrow rings of Uranus. The Saturnian rings, on the other hand, seem to have a distinctly different origin, and are probably formed from material left over during the formation of the planet itself. The gaps that appear in Saturn's rings are not yet fully understood, but the results from Voyager 1 in November 1980 showed them to be far more complex in their structure than had previously been imagined, and their behavior has proved to be extremely puzzling.

So far only Neptune of the major planets has not been shown to be surrounded by a system of rings, but this may only be because the relevant observations are lacking. Neptune is certainly expected to be surrounded by rings, which may have considerable similarities to the Uranian system. Although there is no evidence to suggest that any of the terrestrial planets have ring systems, it has been speculated that the Earth once did; but any such particles would quickly be removed by the sweeping effects of the magnetospheric plasma, and in contrast to the Jovian environment there are no likely sources to replenish the material.

1A

2

1. Jupiter's ring
This composite image of Jupiter's ring (**A**) was made by Voyager 2 on 10 July 1979. The spacecraft was 2° below the plane of the ring, at a distance from Jupiter of 1,550,000 km. A certain degree of blurring occurred, particularly evident in the far right-hand frame, as a result of the motion of the spacecraft. In the detail of the ring (**B**), particles extend almost back to Jupiter; as this high-resolution image suggests, moreover, the ring is made up of several segments. The scratch-like mark is the track made by a star during the exposure.

2. Dark side of Jupiter
The ring showed up particularly brightly when Voyager 2 passed behind Jupiter with respect to the Sun. The planet's shadow obscures part of the ring in the direction of the spacecraft.

3. Planetary ring systems
Jupiter's ring may be compared with those of Saturn and Uranus. It is thought to resemble the rings of Uranus more closely in origin and character, having probably been formed after the planet itself.

3

Map of Io

Volcanoes

Amirani (Plume 5)	27°N, 119°W
Loki (Plume 2)	19°N, 305°W
Marduk (Plume 7)	28°S, 210°W
Masubi (Plume 8)	45°S, 53°W
Maui (Plume 6)	19°N, 122°W
Pele (Plume 1)	19°S, 257°W
Prometheus (Plume 3)	3°S, 153°W
Surt	46°N, 336°W
Volund (Plume 4)	22°N, 177°W

Paterae

Amaterasu Patera	38°N, 307°W
Asha Patera	9°N, 226°W
Atar Patera	30°N, 279°W
Aten Patera	48°S, 311°W
Babbar Patera	40°S, 272°W
Bochica Patera	61°S, 22°W
Creidne Patera	52°S, 345°W
Culann Patera	20°S, 150°W
Daedelus Patera	19°N, 275°W
Dazhbog Patera	54°N, 302°W
Emakong Patera	0°, 110°W
Fuchi Patera	28°N, 328°W
Galai Patera	11°S, 289°W
Gibil Patera	15°S, 295°W
Heno Patera	57°S, 312°W
Hephaestus Patera	2°N, 290°W
Hiruko Patera	65°S, 331°W
Horus Patera	10°S, 340°W
Inti Patera	68°S, 349°W
Kane Patera	48°S, 15°W
Loki Patera	13°N, 310°W
Maasaw Patera	40°S, 341°W
Mafuike Patera	15°S, 261°W
Malik Patera	34°S, 128°W
Manua Patera	35°N, 322°W
Masaya Patera	22°S, 350°W
Maui Patera	20°N, 125°W
Mihr Patera	16°S, 306°W
Nina Patera	40°S, 165°W
Nusku Patera	63°S, 7°W
Nyambe Patera	0°, 345°W
Ra Patera	8°S, 325°W
Reiden Patera	14°S, 236°W
Ruwa Patera	0°, 2°W
Sengen Patera	33°S, 304°W
Shakuru Patera	23°N, 267°W
Shamash Patera	36°S, 152°W
Svarog Patera	48°S, 267°W
Tohil Patera	28°S, 157°W
Ülgen Patera	41°S, 288°W
Uta Patera	35°S, 27°W
Vahagn Patera	27°S, 359°W
Viracocha Patera	62°S, 284°W

Other features

Apis Tholus	11°S, 349°W
Bactria Regio	45°S, 125°W
Chalybes Regio	55°N, 85°W
Colchis Regio	10°N, 170°W
Dodona Planum	60°S, 350°W
Haemus Mons	70°S, 50°W
Inaehus Tholus	29°S, 354°W
Lerna Regio	65°S, 300°W
Mazda Catena	8°S, 315°W
Media Regio	0°, 70°W
Mycenae Regio	35°S, 170°W
Nemea Planum	80°S, 270°W
Silpium Mons	62°S, 282°W
Tarsus Regio	30°S, 55°W

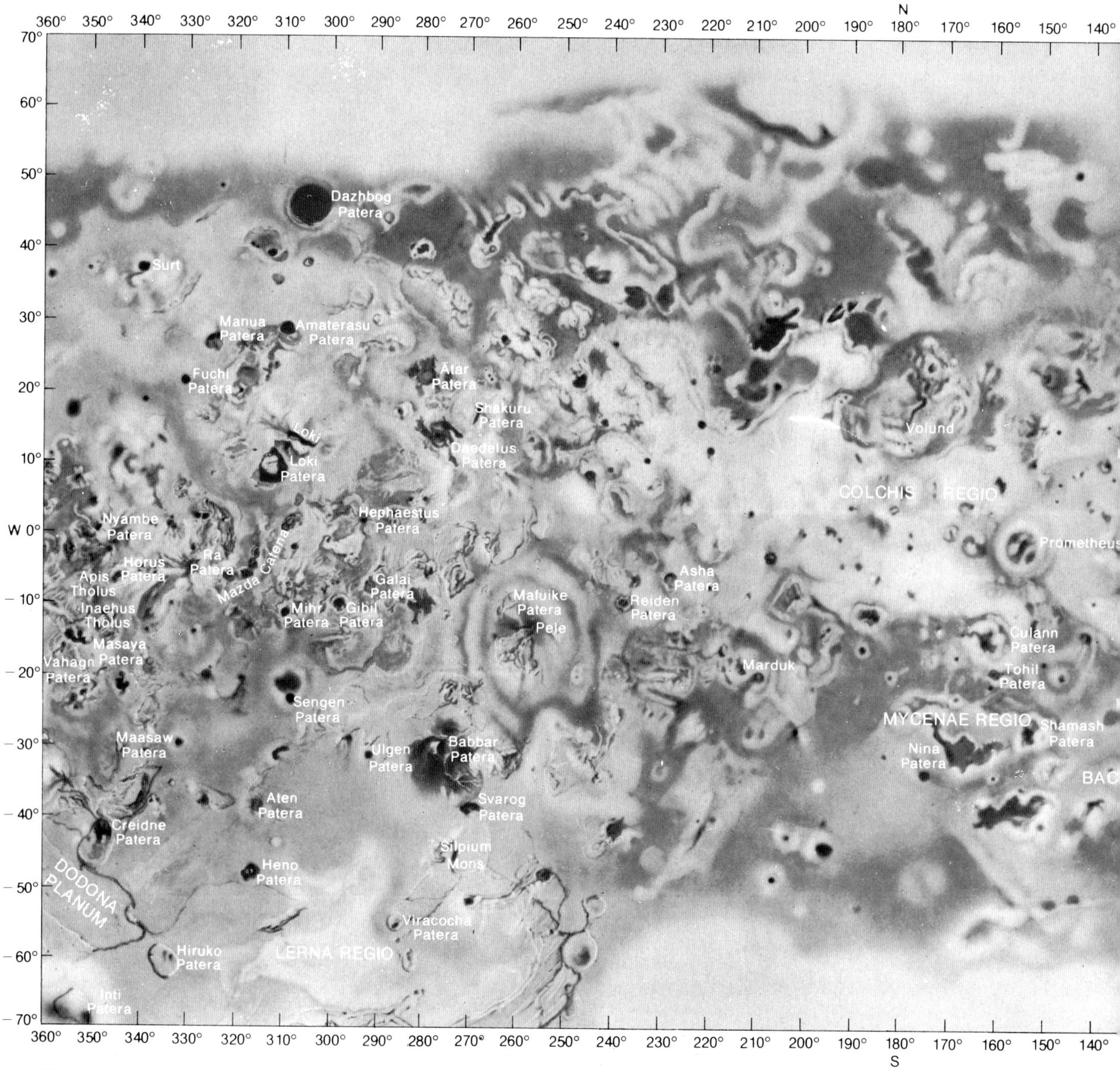

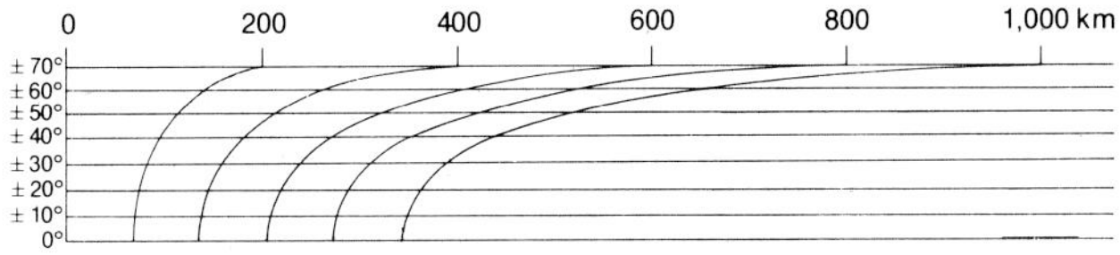

Mercator Projection

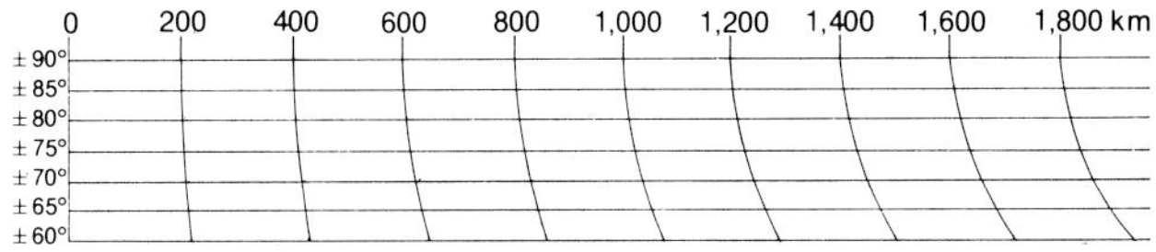

Polar Stereographic Projection

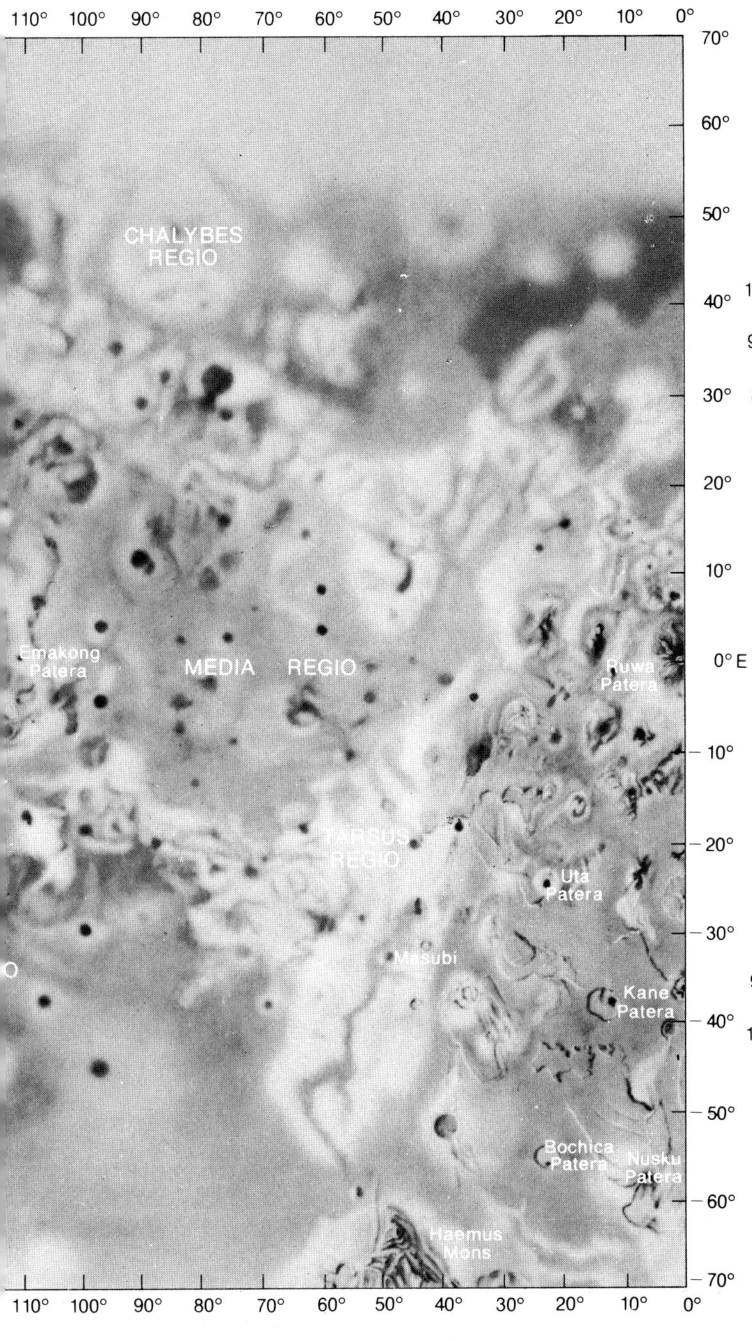

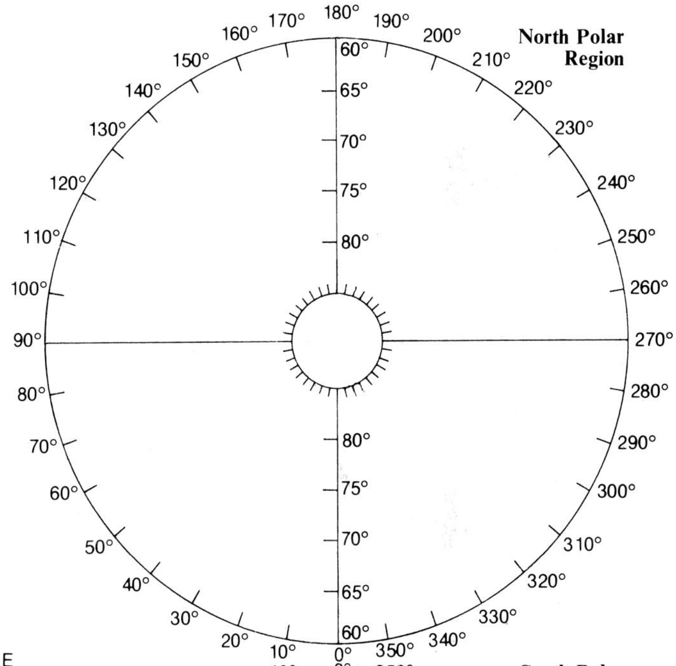

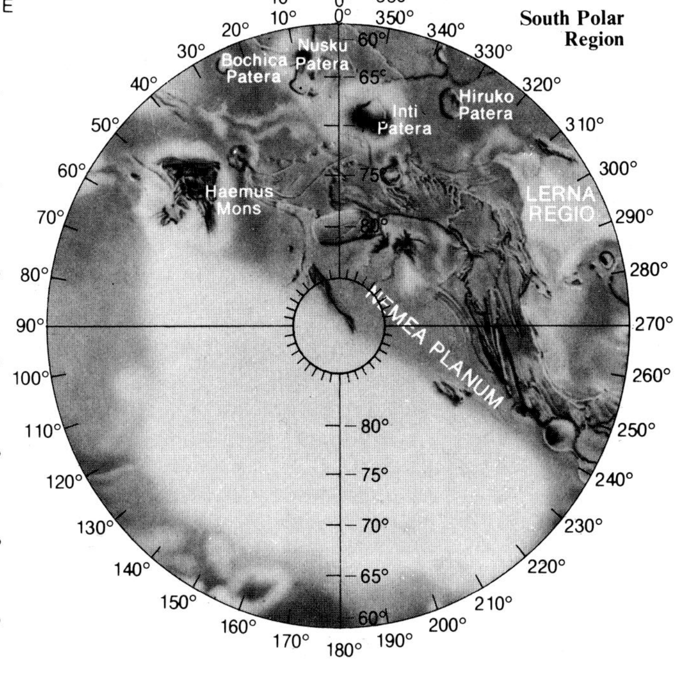

Io

The innermost large satellite, Io, has a radius of 1,820 km and an albedo of 0.63. It is the most dense of the Galileans, with a density of 3.5×10^3 kg m^{-3}, about the same as that of the Moon. Before the Voyager missions it had been assumed that, like the Moon, Io would have a surface covered with craters, but in the event this view proved mistaken. Shortly before Voyager 1 made its closest approach to Io, however, it had been suggested that as the satellite was subjected to the gravitational attraction of Jupiter on the one hand and of the other Galileans (particularly Europa) on the other, its surface would be flexed and the resulting friction could produce enough heat to cause a molten interior, with heat being released through the crust in the form of volcanic eruptions.

This prediction was confirmed dramatically on 9 March 1979. At the Jet Propulsion Laboratory a check was being made of the orbital position of Io by measuring its position relative to a faint star, known by its catalogue number of AGK-10021. When the star appeared on the screen of the imaging module, so did a huge umbrella-shaped plume at the edge of Io. This could be nothing but an active volcano, with a plume rising to some 280 km above the Ionian surface.

Pictures from Voyager 1 (which passed Io at a distance of 420,100 km) revealed a surface strikingly different from the lunar appearance that had been expected. The color was red and orange, and the dominant features were violently active volcanoes. Eight eruptions were seen, and the ejection velocities were calculated to be over 1 km s^{-1}, more violent than Etna, Vesuvius or even Krakatoa. The explosive plumes were from 70 to 300 km high, with the material sometimes rising to around 100 km and fanning out to form an umbrella-shaped cloud before falling back to the surface. When Voyager 2 passed at a distance of 558,270 km, the general appearance of the scene was much the same, though there had been changes in detail.

According to present ideas of Io's internal structure, there is a sulphur and sulphur dioxide crust overlying a molten silicate interior, while the actual core may well be solid. Various theories have been proposed to explain the volcanism. For instance, it has been suggested that a crust about 20 km deep is made to rise and fall regularly by 100 km or so by tidal forces, and that heat, produced by the friction, then escapes through the surface vents as volcanic eruptions. In the "silicate-volcanism" model, silicon-enriched magma erupts through a silicate crust which is rich in sulphur. The general process could not be unlike that of terrestrial volcanism, allowing for the greater abundance of sulphur on Io; variations in the chemistry and physical conditions of the magma chamber could account for the fact that not all eruptions are of the same type.

Alternatively, there may be a "sea" of sulphur and sulphur dioxide about 4 km deep, with only the uppermost kilometer frozen; this ocean has been forced above the silicate sub-crust because of thousands of millions of years of tidal heating. Violent volcanism occurs in the crust when liquid sulphur dioxide meets molten sulphur and explodes into space as it is decompressed. Sulphur seeping back into the molten silicate interior would then balance the extra sulphur deposited on the surface.

Io's low escape velocity prevents the satellite from retaining an atmosphere of appreciable density. However, even before the Pioneer mission, some astronomers had reported a temporary increase in brightness of about a tenth of a magnitude immediately after Io emerged from an eclipse, and it was suggested that reflective material had been deposited on the surface while the sunlight was cut off, only to evaporate again as soon as the eclipse finished.

Then, in December 1973, Pioneer 10 detected an "ionosphere". Moreover, Earth-based observations also revealed an electrically neutral atmosphere with a surface pressure of below 10^{-3} Pascals (*see* pages 22–23). It was suggested that the surface of Io could be covered with salt, in which case this "sodium cloud" would be produced by the effects of Jovian radiation on the salty covering.

1

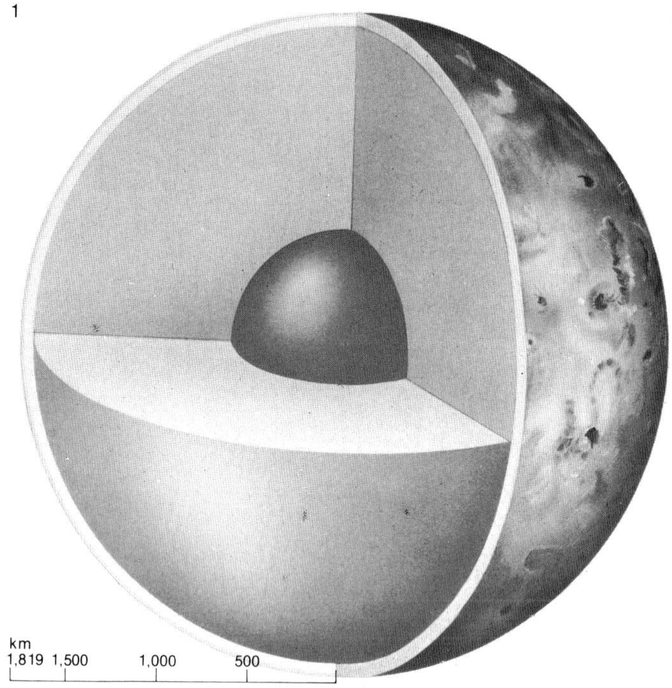

1. Interior of Io
Io's density has been determined more accurately than those of the other Galileans; this knowledge is helpful in constructing models of the interior structure. Io is known to be rich in sulphur; its crust is probably composed of sulphur and sulphur dioxide, while the interior is thought to consist of molten silicates surrounding a solid core.

2

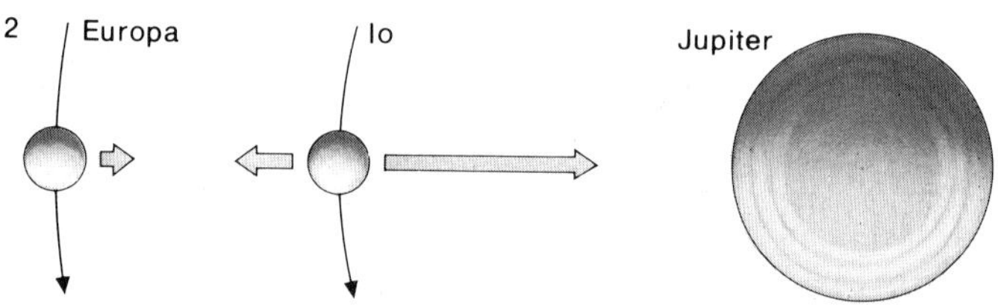

2. Gravitational effects
Io is subject to the gravitational pull not only of Jupiter but also of the other Galileans, particularly Europa. When Io passes between Jupiter and Europa, the opposing forces on it cause its surface to flex, and the resulting internal friction produces heat which may be responsible for the volcanic activity.

3. Volcanism on Io
The first evidence of volcanic activity on Io came from this image, when the curve on Io's limb was identified as a cloud of ash thrown up by an erupting volcano. A second eruption—the bright spot on the terminator—can also be seen. The image was made by Voyager 1 on 8 March 1979.

3

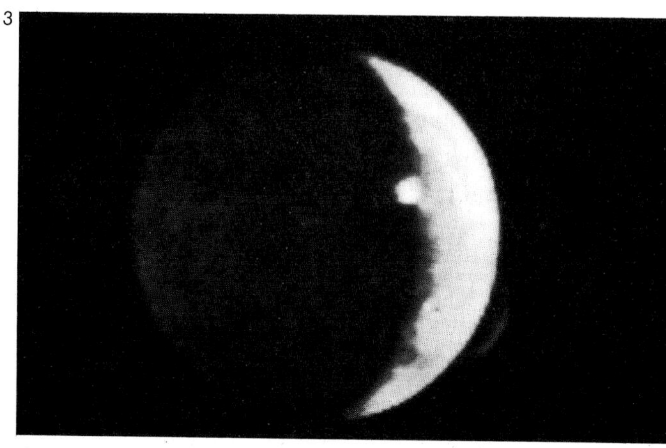

4A

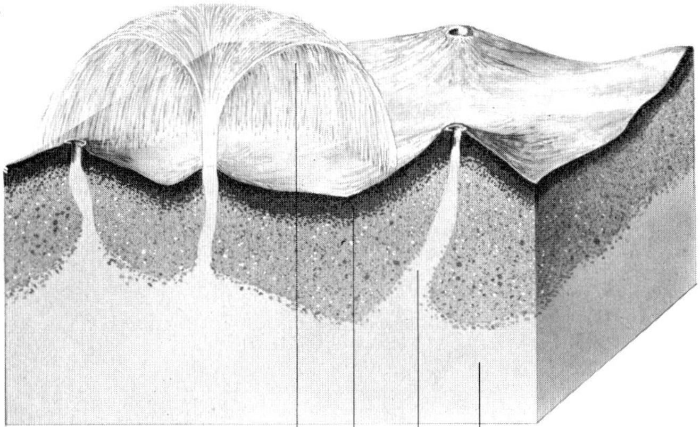

B

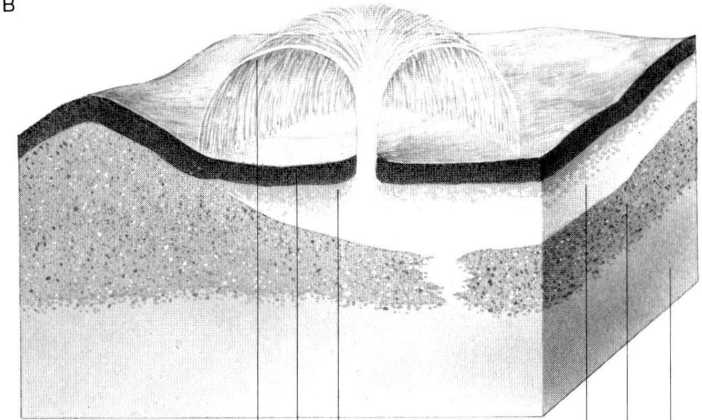

4. Causes of volcanism
Several explanations of Io's volcanic activity have been proposed. The "silicate-volcanism" model (**A**) involves a mechanism similar to the volcanic process on Earth. Silicate magma is forced up through the crust, erupting through vents in the surface. The magma is richer in sulphur than its terrestrial equivalent. An alternative model (**B**) attributes the volcanic explosions to the effect of molten sulphur coming into contact with liquid sulphur dioxide. The sulphur dioxide is suddenly decompressed and explodes to form plumes. According to this model "oceans" of molten sulphur lie close beneath the topmost crust, having been forced upwards through the molten silicate interior. These reservoirs are continually replenished by sulphur seeping back through the surface as the ejected material falls back to the ground.

5

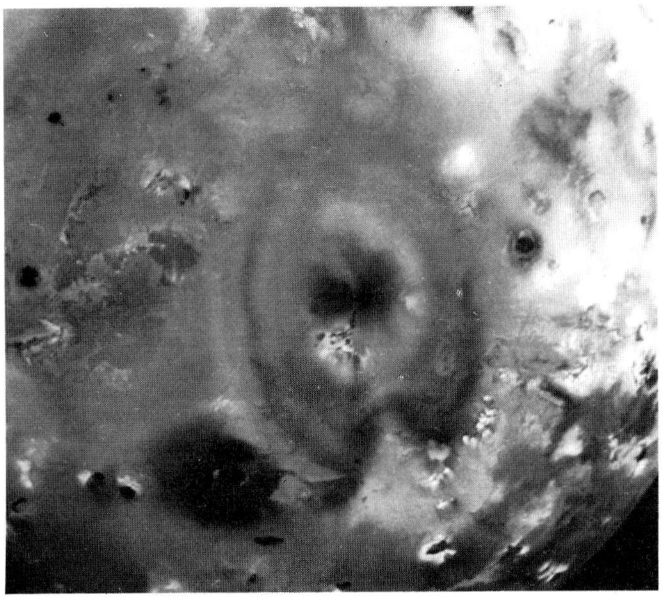

5. Pele (19°S, 257°W)
The source of the volcanic cloud known as Plume 1 was a heart-shaped volcano, now named Pele. Its diameter is about 1,200 km at the widest point.

6. Loki (19°N, 305°W)
The surface markings in the region of the eruption known as Plume 2, whose source has now been given the name Loki, altered significantly between the encounters of Voyager 1 (**A**) and Voyager 2 (**B**). Volcanic ejecta appear to have obscured older patterns, and bright deposits are visible in several places.

7. Maasaw Patera (40°S, 341°W)
The dark features are lava flows emanating from a caldera.

8. Gibil (15°S, 295°W)
This image shows one of the named volcanic features. Dark lava flows can be seen in the upper left corner.

6A

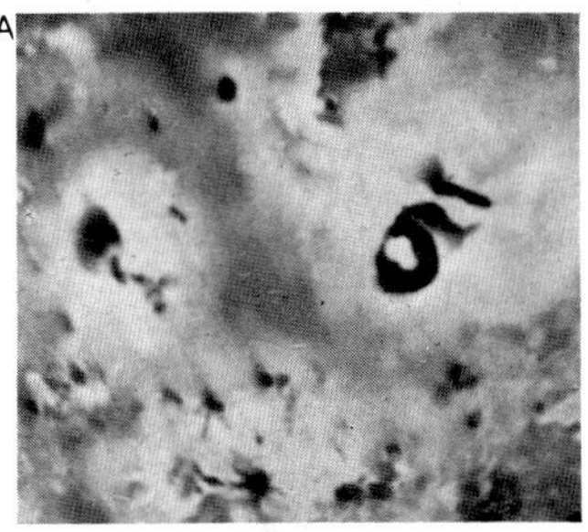

B

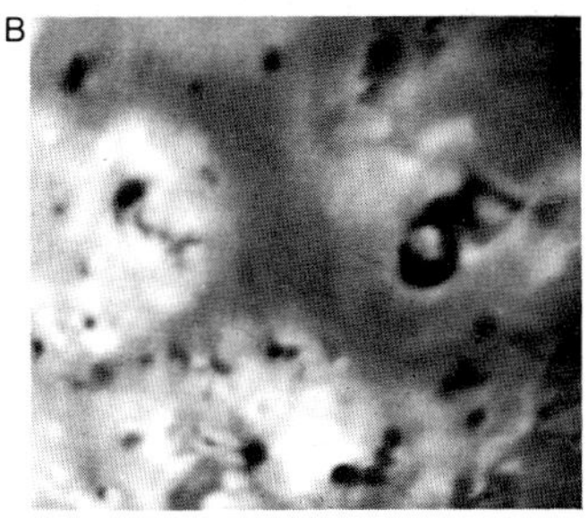

7

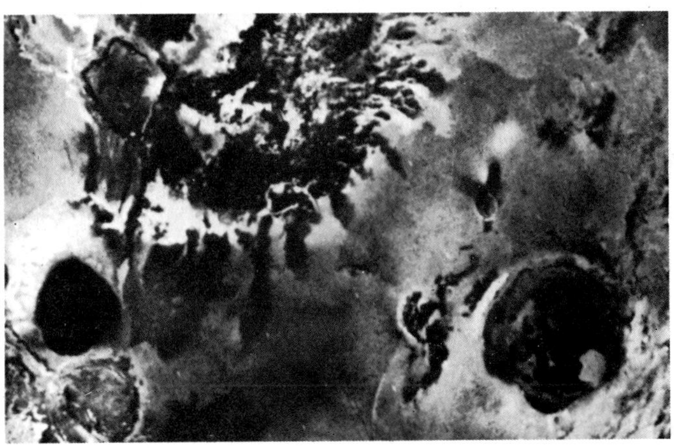

8

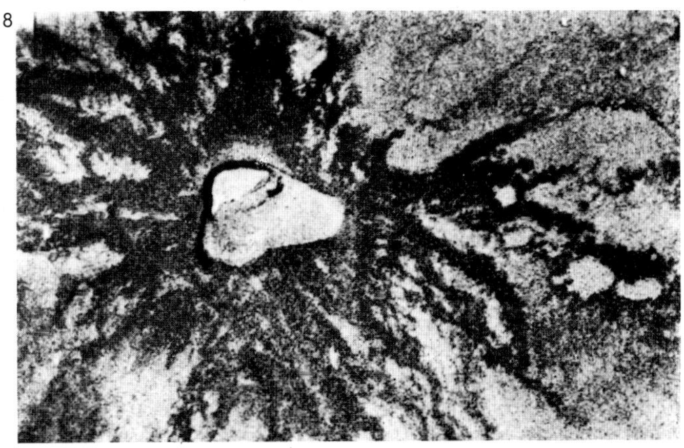

Io Photomosaics I

West region
The surface of Io is active, and so the finer details are subject to change even over short periods of time, but the main features are presumably permanent. The most significant formations are the volcanoes, such as Pele and Loki. Pele was active during the Voyager 1 pass, but inactive when Voyager 2 made its flyby; not far from it is the dark area of Babbar Patera. Loki was active during both passes, and seems to be one of the most violent of the Ionian volcanoes; between it and Pele is the prominent feature known as Hephaestus Patera. Galai and Gibil are other notable objects. The region to the south of Pele and Babbar Patera—the Lerna Regio—has fewer well-defined features and is lighter in color. Ra Patera is another highly volcanic area, with Mazda Catena running alongside it. Also prominent is the structure of Haemus Mons. (The globe appears distorted because different areas are shown from slightly different viewpoints.)

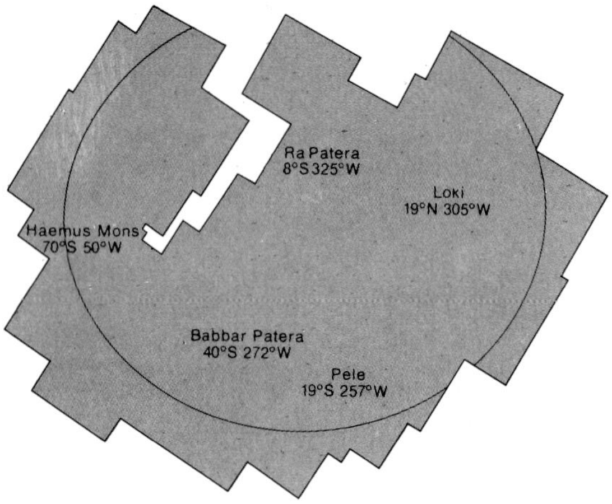

Ra Patera
Mazda Catena
Loki
Hephaestus Patera
Gibil Patera
Galai Patera
Babbar Patera
Pele

Io Photomosaics II

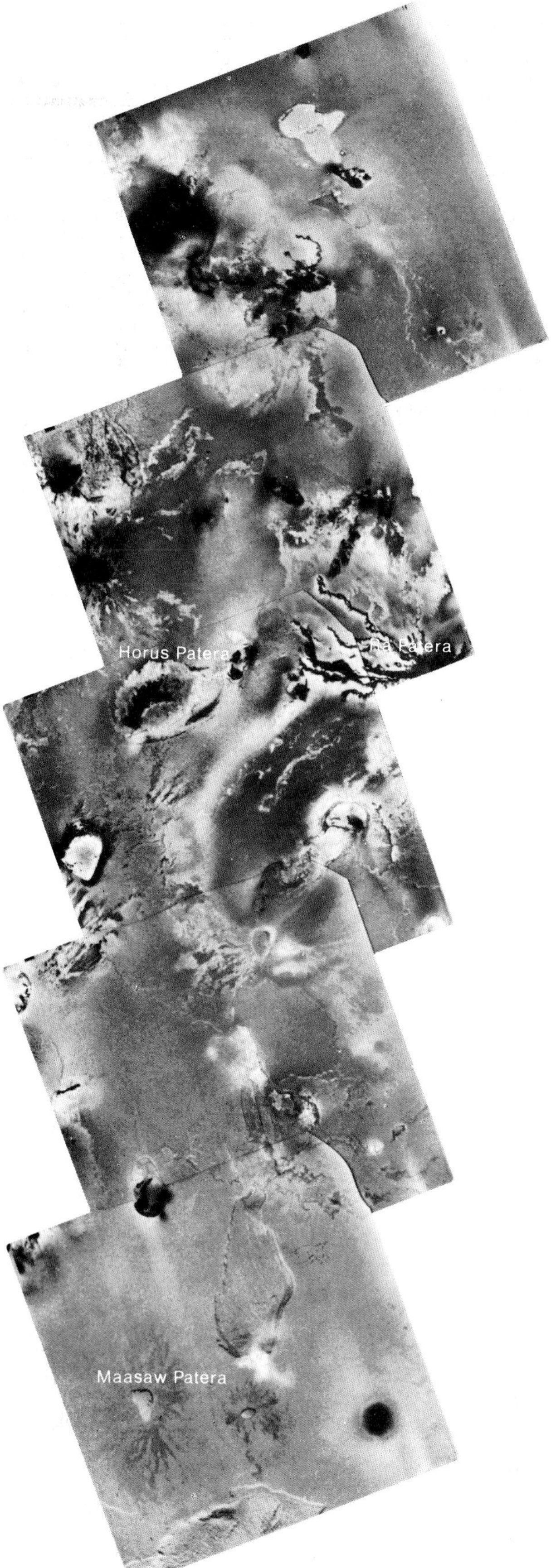

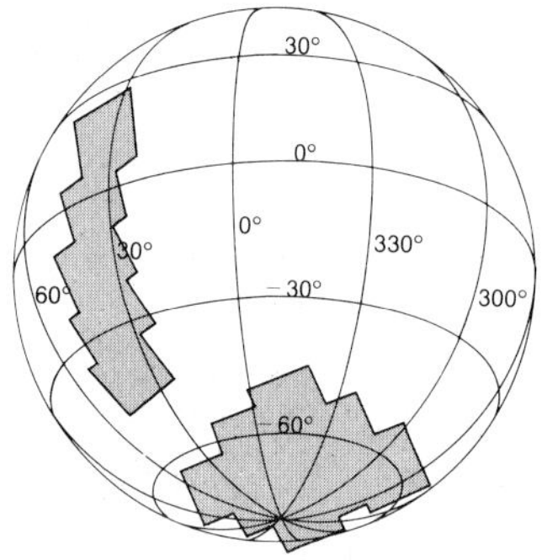

1. West region
A complex, volcanic area; Ra Patera. Maasaw Patera has a distinctive heart shape.

2. South polar region
Part of Lerna Regio is shown, with the prominent Haemus Mons, Inti Patera and Nusku Patera; Dodona Planum is relatively lacking in detail.

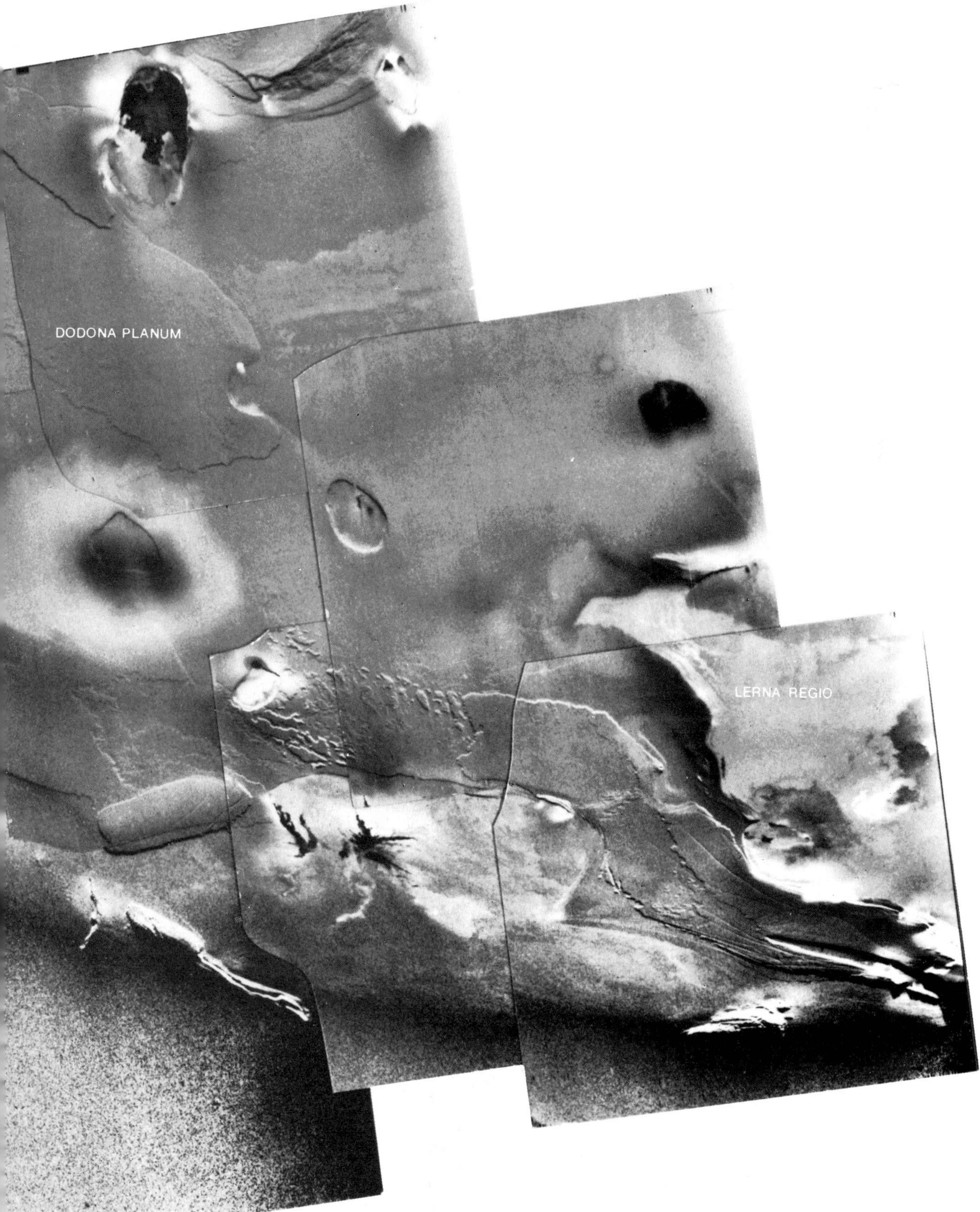
DODONA PLANUM
LERNA REGIO

Map of Europa

Adonis Linea
Agenor Linea
Argiope Linea
Asterius Linea
Belus Linea
Cadmus Linea
Libya Linea
Minos Linea
Pelorus Linea
Phineus Linea
Sarpedon Linea
Thasus Linea

Cilicia Flexus
Gortyna Flexus
Sidon Flexus

Thera Macula	45°S, 178°W
Thrace Macula	44°S, 169°W
Tyre Macula	24°N, 144°W

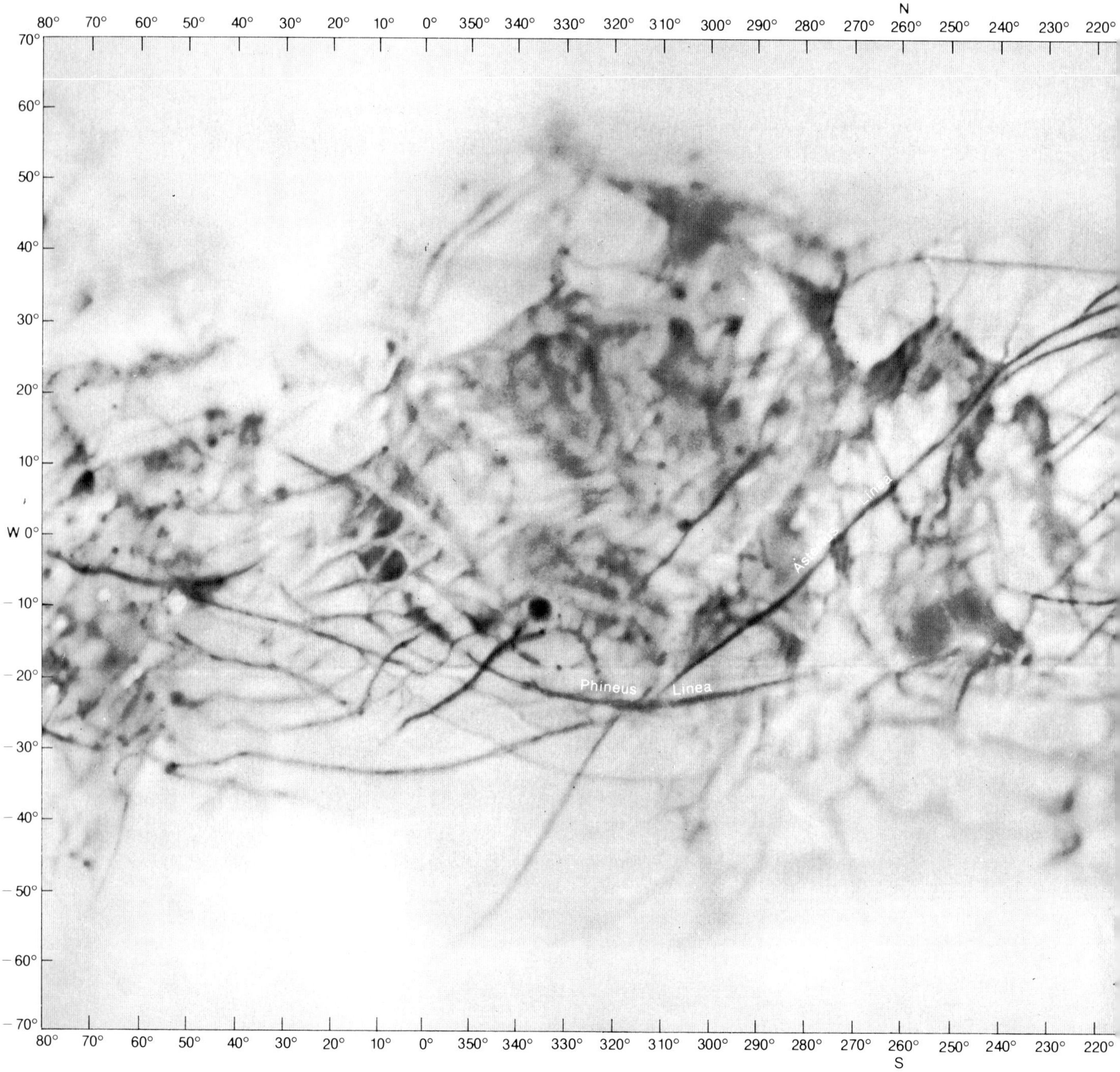

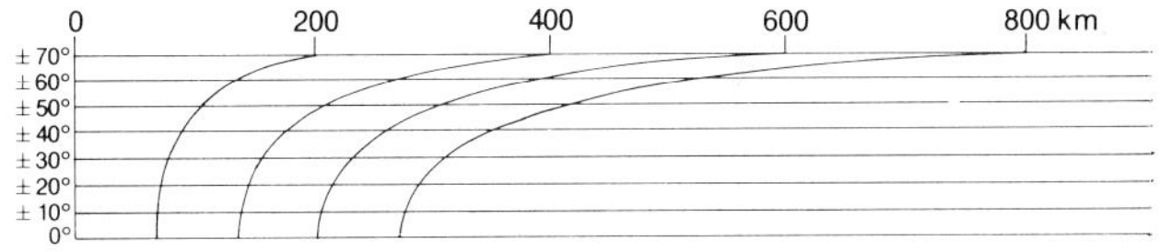

Mercator Projection

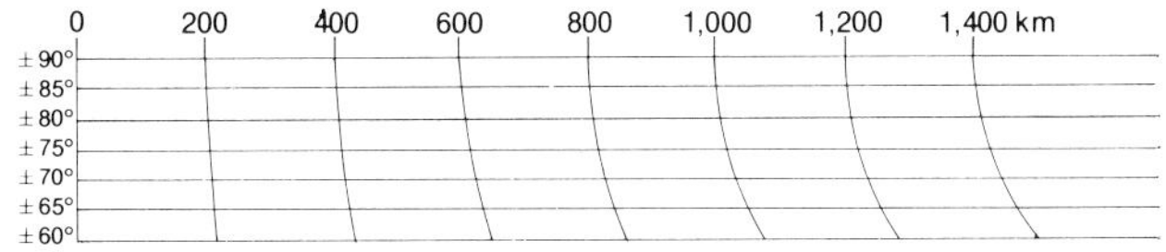

Polar Stereographic Projection

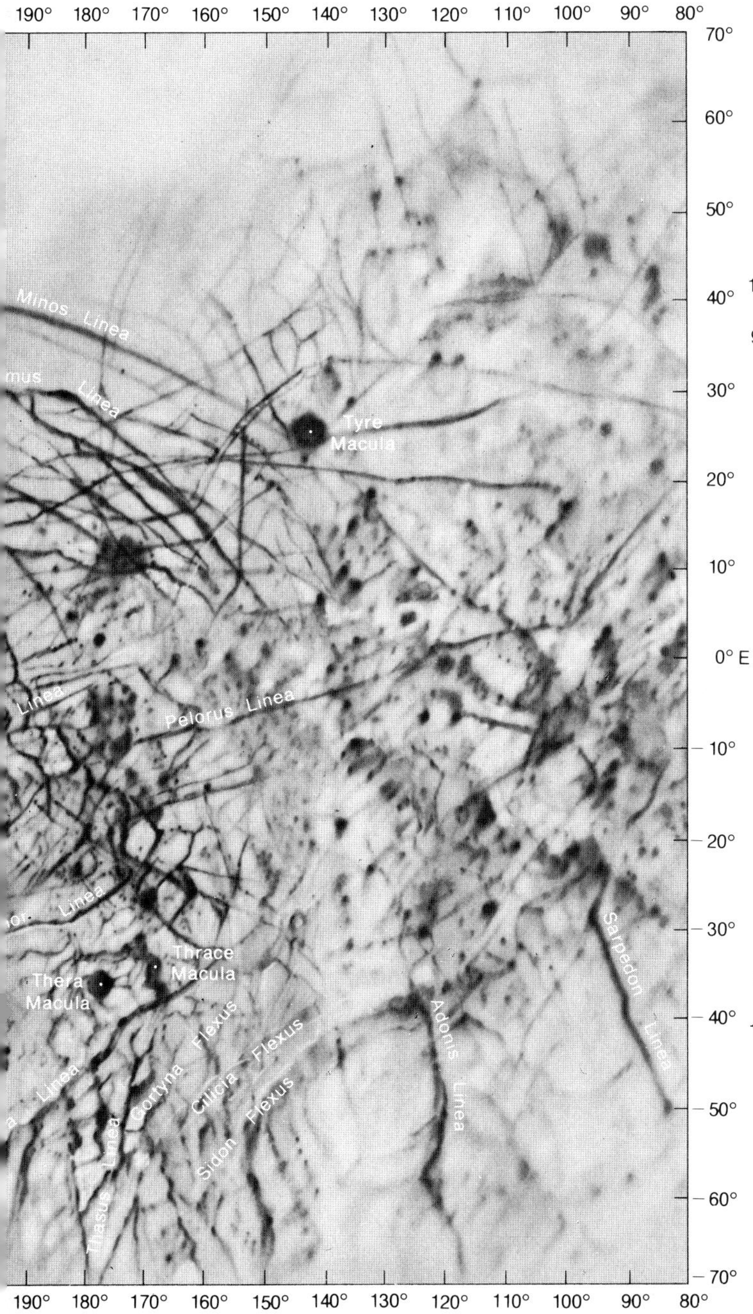

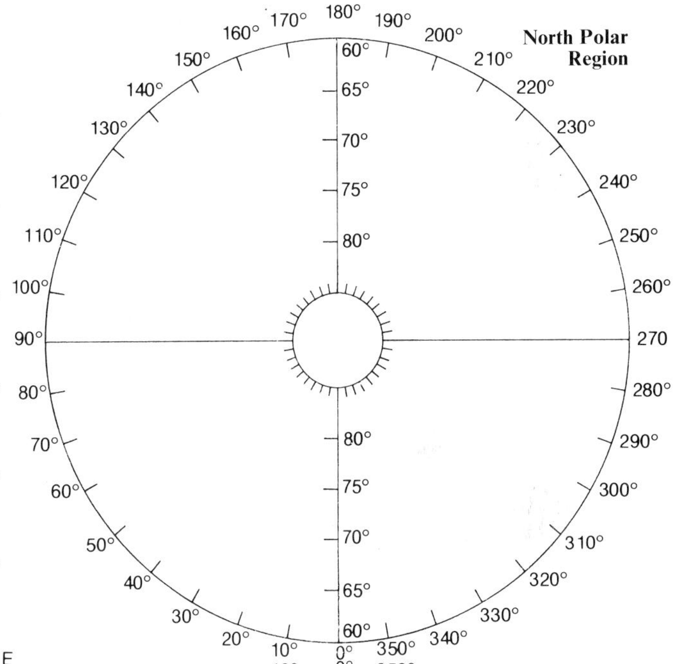

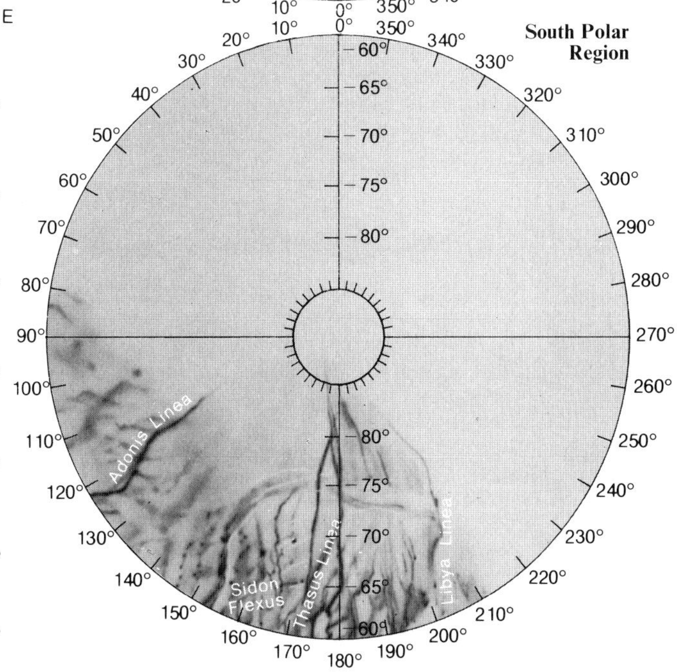

Europa

With a radius of 1,525 km, Europa is the smallest of the Galilean satellites, and the only one inferior in size and mass to the Moon. The overall density is about 3.3×10^3 kg m^{-3}, appreciably less than that of Io, but still great enough to indicate a relatively large silicate core. Europa is the most reflective of the Galileans and has an albedo of 0.64, but in the days before the Voyager mission there was no reason to suppose that it was markedly different from Io. However, although Voyager 1 did not approach Europa closer than 732,230 km and although the resolution of the pictures of its surface was much lower than with the other Galileans, they showed Europa to be totally unlike Io. The general hue was whitish instead of red; there were no volcanoes, active or dead, and there were practically no craters. Even more strange was the apparent absence of vertical relief, as revealed by pictures of the satellite's terminator. Europa is remarkably smooth, to the extent that it has been compared with a billiard-ball.

Voyager 2 made a much closer approach, 204,030 km at its nearest point, and sent back high-quality pictures which revealed darkish, mottled regions and lighter areas; but the main impression was of a complex maze of bright and dark lines, criss-crossing each other. There were linear features up to 40 km wide and thousands of kilometers long, with narrow ridges having widths of up to 10 km and lengths of at least 100 km.

New terms have been introduced to describe Europa's terrain: "linea" (a dark or bright elongated marking, either straight or curved), "flexus" (a very low curvilinear ridge with a scalloped pattern), and "macula" (a dark spot, sometimes irregular in shape). Yet no particular feature stands out prominently to compare with those on the other Galileans. As far as Europa is concerned one feature is remarkably like another. The three named "macula" objects (Thera Macula, Thrace Macula and Tyre Macula) can be identified without difficulty, but are not very striking.

Craters

One of the really puzzling aspects is the paucity of craters. On Io any craters would soon be obliterated by the constant surface activity, but Europa is quiescent, and there is little doubt that the visible surface is made up of ice. The three craters identified with fair certainty are between 18 and 25 km in diameter, but they are not alike in morphology, one is bowl-shaped, another shallow and associated with what seems to be a system of dark rays, and the third is raised up, as though the surrounding surface had subsided for some reason or other.

One reasonable suggestion was that water had surged upward to the surface, forming a layer of ice over the silicate interior. If the layer were as much as 100 km thick, it would be more than adequate to cover any surface relief; but it may be much thinner than this, so that in places the silicate base is not far below the visible surface. It is also possible that there is a relatively thin ice crust lying over water, or softer ice which could be described as "slush". However, the surface of Europa is at least several hundreds of millions of years old and more probably several thousands of millions. The lack of impact craters still has to be accounted for, assuming that craters of the type found on Ganymede and Callisto really have been produced by meteoritic bombardment.

One possible answer is that the crust remained comparatively warm and slushy until the main bombardment era was over. Radioactive heating could have been involved, and Europa must also experience interior tidal forces which heat it in much the same way as with Io, although since Europa is further from Jupiter the effect is only about 10 percent as great. Assuming that the outer crust is rigid, the fractures have presumably been filled in with material rising from below. It seems that after fracture the pieces of crust remained in their original positions, in contrast to Ganymede, where fragments of crust appear to have shifted about relative to each other.

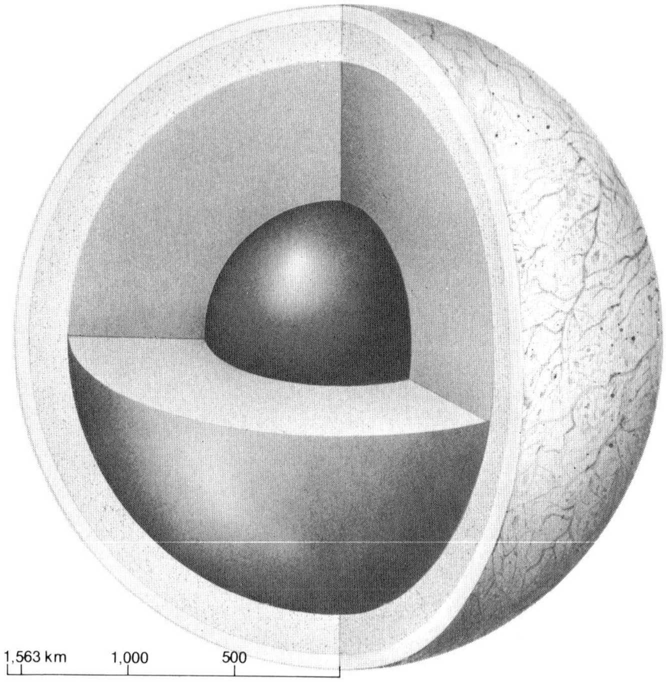

1. Interior of Europa
Europa is believed to have a relatively large silicate core. Its high albedo suggests that it has an icy surface. Europa's density of 3.1×10^3 kg m^{-3} would result from a mixture of silicates and water-ice in the ratio of 9:1 by weight. A 70 km thick crust of ice may cover a region of slushy water-ice, 100 km thick, beneath which lies the 1,400 km silicate core. Radioactive decay is likely to continue to heat the interior, which may reach temperatures of around 2,800 K.

2. Terminator
Europa's terminator shows the surface to be extremely smooth. The absence of impact craters suggests that the surface may have been too soft to have retained impressions made by meteoritic bombardment. This image, made by Voyager 2 on 9 July 1979, reveals bright ridge-like features rising to about 100 m above the surface. They are typically 100 km long and between 5 and 10 km in width. Broader dark bands are also apparent. These may be several thousand kilometers long.

2
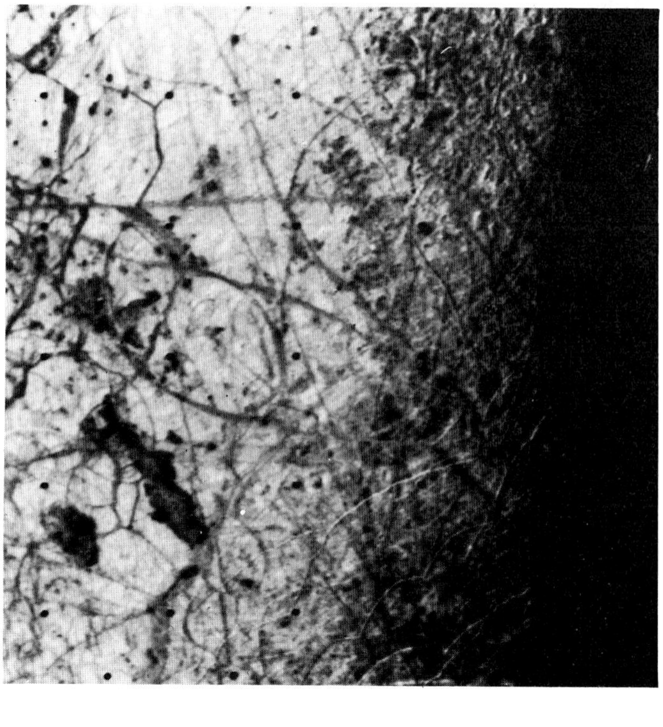

3

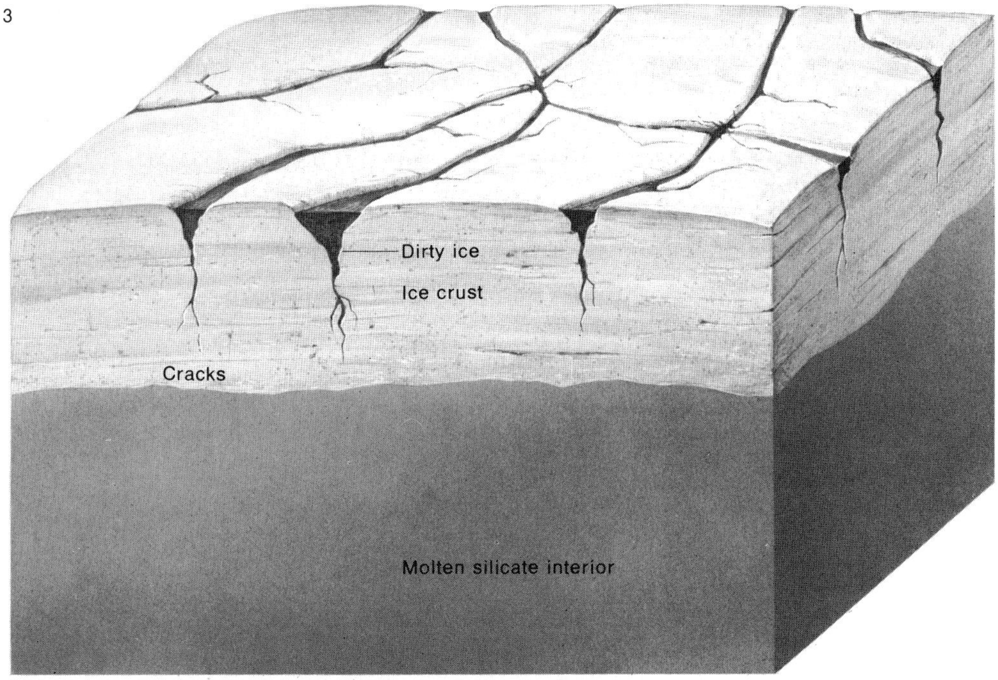

3. Surface structure
The markings on Europa's surface suggest that fractures have occurred in the crust and that material from below has been forced upwards to fill the cracks. It is possible that at present there is a 100 km thick layer of water or water-ice separating the crust from the silicate interior, but this layer is not shown in the illustration as considerable uncertainty exists.

4

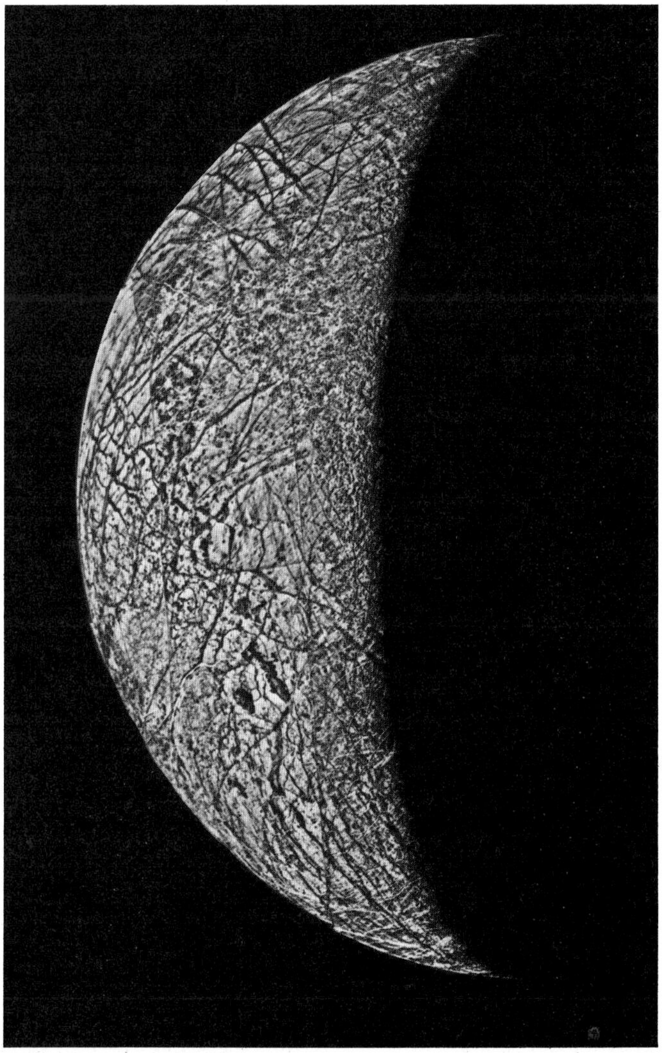

4. Mosaic of Europa
Two basic types of terrain are apparent in Europa's equatorial region, one bright, the other mottled and slightly darker. The darker terrain exhibits small depressions which may, in fact, be impact craters.

5. Cadmus Linea and Belus Linea
Two of the named "linea" features are shown in this image. Cadmus Linea is the uppermost dark band in the picture, while Belus Linea is the lower of the two nearly parallel bands cutting diagonally across the image from lower left to upper right. The two features intersect at 27°N, 172°W.

6. Tyre Macula (34°N, 144°W)
This image shows one of the dark, irregular-shaped features that have been named. (The others are Thera Macula and Thrace Macula.)

5

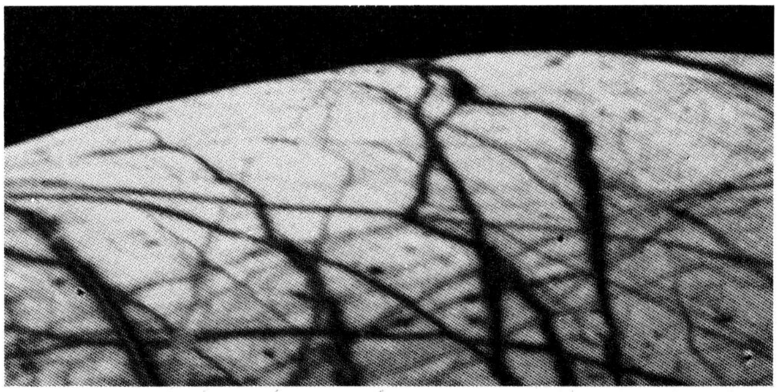

6

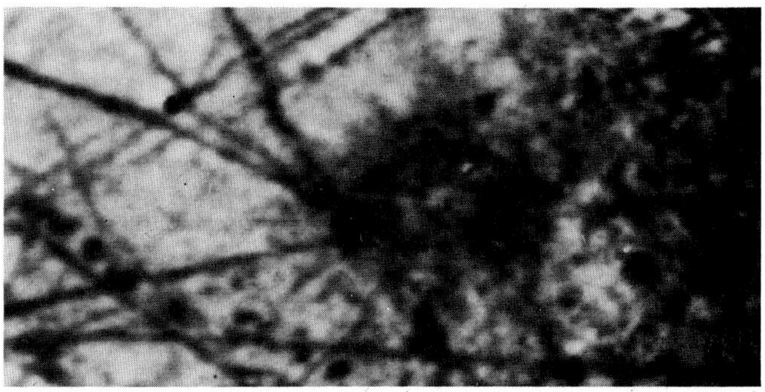

Map of Ganymede

Craters	
Achelous	66°N, 4°W
Adad	62°N, 352°W
Adapa	83°N, 22°W
Ammura	36°N, 337°W
Anu	68°N, 332°W
Asshur	56°N, 325°W
Aya	67°N, 303°W
Ba'al	29°N, 326°W
Danel	4°N, 21°W
Diment	29°N, 346°W
Enlil	52°N, 301°W
Eshmun	22°S, 187°W
Etana	78°N, 310°W
Gilgamesh	58°S, 124°W
Gula	68°N, 1°W
Hathor	70°S, 265°W
Isis	64°S, 197°W
Keret	22°N, 34°W
Khumbam	15°S, 332°W
Kishar	78°N, 330°W
Melkart	13°S, 182°W
Mor	35°N, 323°W
Nabu	36°S, 2°W
Namtar	49°S, 343°W
Nigitsu	48°S, 308°W
Nut	61°S, 268°W
Osiris	39°S, 161°W
Ruti	15°N, 304°W
Sapas	59°N, 31°W
Sebek	65°N, 348°W
Sin	56°N, 349°W
Tanit	59°N, 32°W
Teshub	2°N, 16°W
Tros	20°N, 28°W
Zaqar	60°N, 31°W

Regiones	
Barnard Regio	22°N, 10°W
Galileo Regio	35°N, 145°W
Marius Regio	10°S, 200°W
Nicholson Regio	20°S, 0°W
Perrine Regio	40°N, 30°W

Sulci	
Anshar Sulcus	15°N, 200°W
Apsu Sulci	40°S, 230°W
Aquarius Sulcus	50°N, 10°W
Dardanus Sulcus	20°S, 13°W
Harpagia Sulci	0°, 317°W
Kishar Sulcus	15°S, 220°W
Mashu Sulcus	22°N, 200°W
Mysia Sulci	10°N, 340°W
Nun Sulci	50°N, 320°W
Philus Sulci	37°N, 215°W
Phrygia Sulcus	20°N, 5°W
Sicyon Sulcus	44°N, 3°W
Tiamat Sulcus	3°S, 210°W
Uruk Sulcus	0°, 157°W

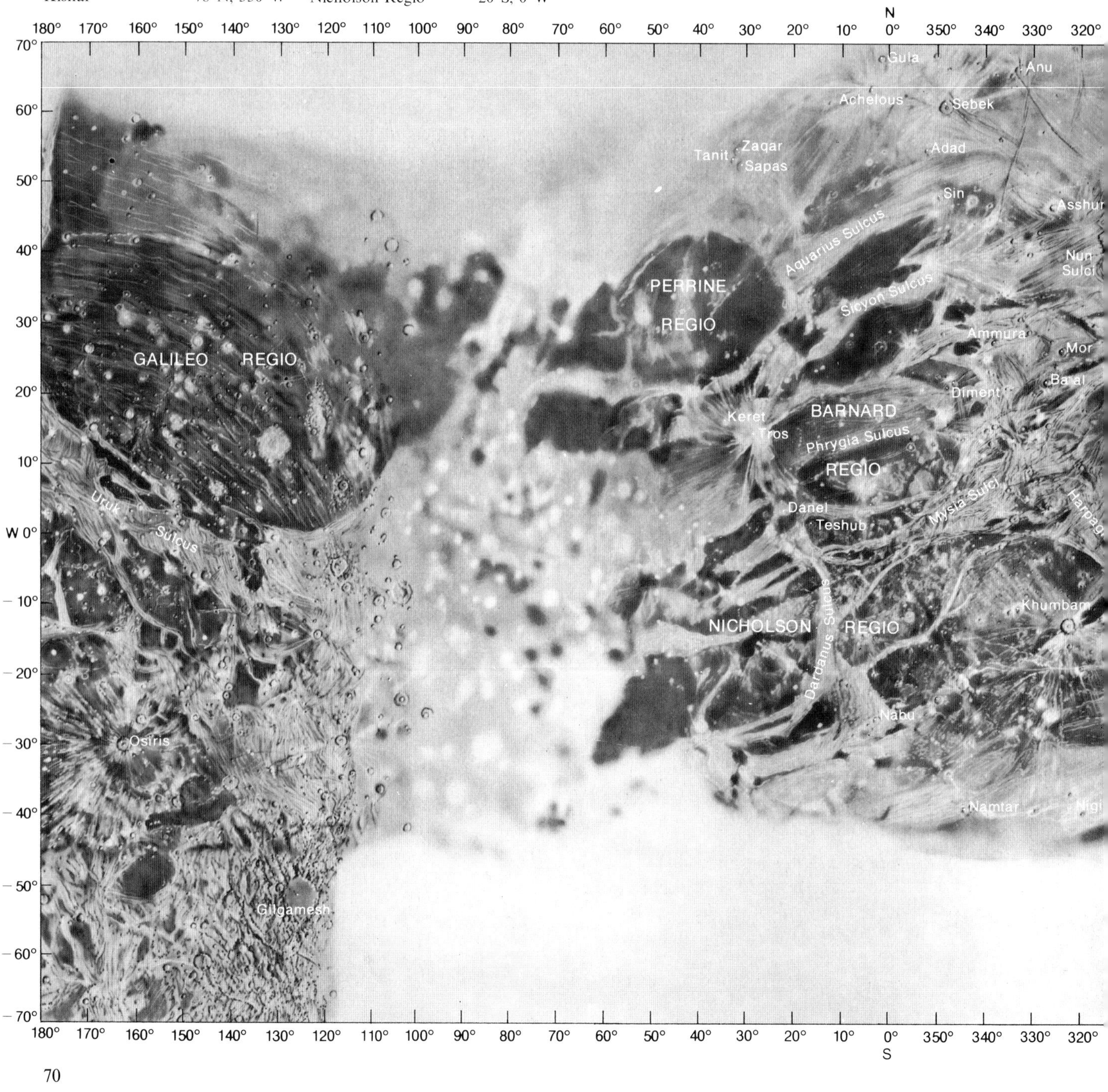

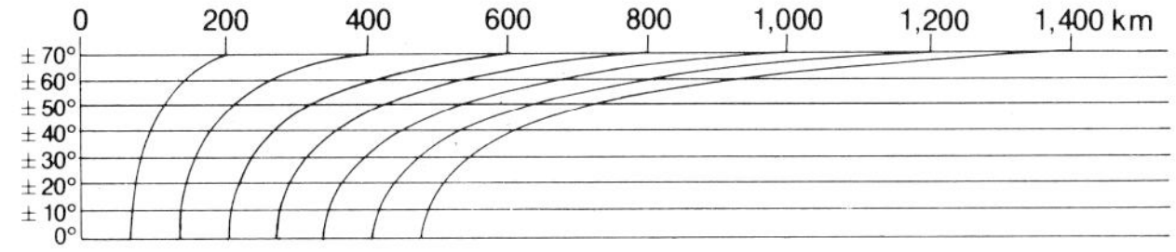

Mercator Projection

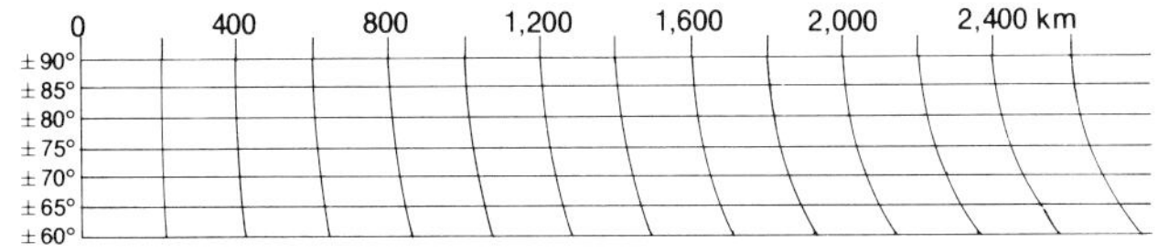

Polar Stereographic Projection

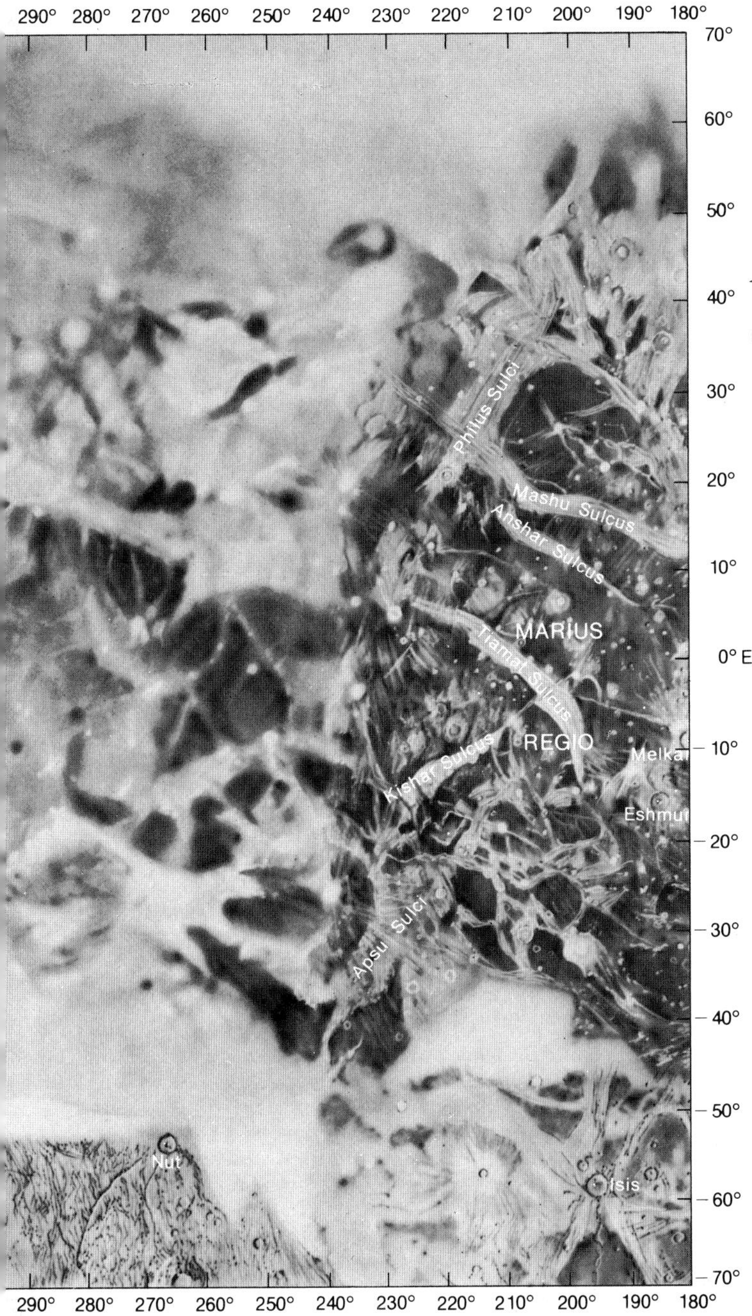

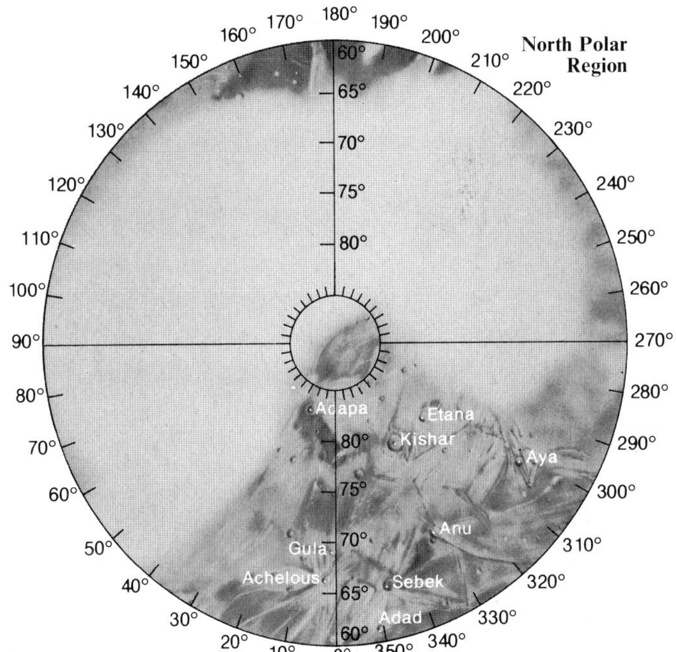

North Polar Region

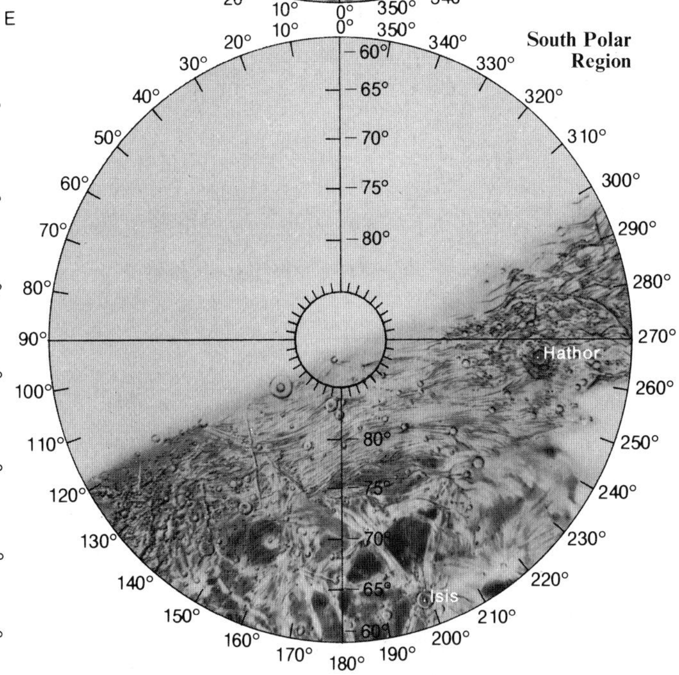

South Polar Region

Ganymede

Ganymede, the third of the principal satellites, is the largest and brightest member of the Jovian family. It has an albedo of 0.4, and a radius of 2,635 km, and is, in fact, one of the largest satellites in the Solar System, rivalled only by Titan in Saturn's system and Triton in Neptune's. Telescopically, Ganymede is always easy to identify, and observers using large telescopes have been able to make out at least one surface feature, the huge circular, darkish region shown in detail by the Voyagers and now named Galileo. The overall density of Ganymede is less than twice that of water, and from this it may be inferred that it consists of about half rock and half water (or ice). The crust is presumably composed of ice mixed with rocky materials, and is darkest in the most ancient regions. It is thought to cover a convective layer of water or soft ice, which in turn surrounds a small silicate core. It is very unlikely that major changes take place there now. There may be a certain amount of radioactive heating, and present-day tidal effects must also contribute to the heating of the satellite. Ganymede moves inside Jupiter's magnetosphere. No trace of atmosphere has yet been detected.

Surface features

Both Voyagers obtained detailed views of Ganymede. The first probe passed at a distance of 112,030 km, and the second at only 59,530 km. Eighty percent of the total surface has been covered down to a resolution of 5 km or less.

The first impression is of a surface superficially similar to that of the Moon, with craters of various kinds (some of them with central peaks), ray systems, and brighter regions which are presumably younger than the dark terrain. If the craters are mainly of impact origin, it would lead to the assumption that Ganymede was subjected to very intense bombardment in the distant past. The surface features are certainly very old compared with those of Europa, and even more so compared with Io. Yet in some respects the surface is not truly "lunar". The relatively smooth terminator shows that the surface relief is much less marked than that of the Moon, and this is only to be expected if the materials there are largely icy.

Generally speaking there are two main types of geological unit. First there is the ancient, darkish, heavily cratered terrain. Such regions are often roughly polygonal, and may be up to several tens of kilometers across. Secondly there are brighter, presumably younger regions which are characterized by "bundles" of long, parallel grooves. The term "sulcus", meaning a groove or furrow, has been used to describe these features. However, close studies of the Voyager results show that these two types of terrain intermingle, so that the overall picture is one of great complexity. The large stripe-like features are characteristic of Ganymede, and it is these which divide the cratered terrain into isolated polygons up to a thousand kilometers across.

It is likely that Ganymede's surface was originally darkish and composed of soft ice, and that it cooled and froze rapidly during the first few hundred million years after Ganymede was formed. The grooved terrain developed gradually, and the grooves were filled with frozen water; faults are common. (Ganymede is incidentally the only world, apart from the Earth, known to exhibit lateral faulting.) Some of the craters are ghostly, reminiscent of Stadius on the Moon, but others look comparatively fresh, and there are ray systems which undoubtedly represent the youngest of all the features. Around these ray craters the ice is almost white, because it is so much "cleaner" than the rest of the surface.

It is probable that the surface is incapable of supporting much weight. There must be more water than on Europa, and the grooves have been interpreted as valleys between ice extrusions. It has been suggested that the past surface activity, great enough to create "tectonic blocks" (icy versions of the Earth's plates), was due to the slowing-down of Ganymede's axial rotation by the tidal forces that produced the present synchronous orbit.

1

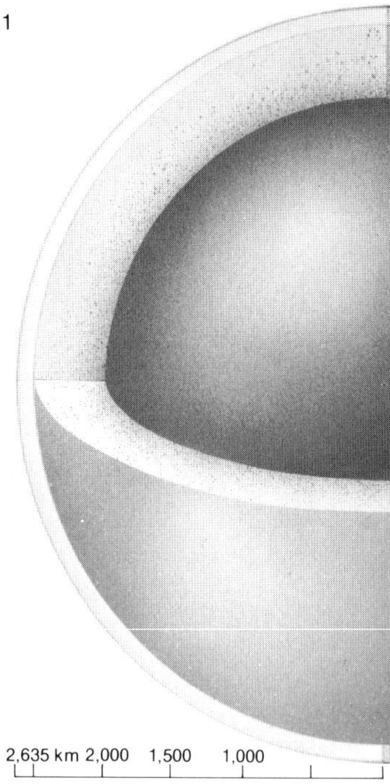

1. Interior of Ganymede
Ganymede is believed to have an ice crust less than 100 km thick, with a convecting mantle of water or soft ice between 400 and 800 km thick. Below this there is thought to be a relatively large silicate core of radius 1,800 to 2,200 km. The icy surface has become dirty with age, except where fresh white ice has been ejected as a result of meteoritic impacts. At some stage in its history, Ganymede's surface consisted of "tectonic" blocks of ice which shifted about relative to one another, rather like the Earth's continental plates.

2. Terminator
The surface of Ganymede is heavily cratered and scored with a large number of grooves. These are typically between 5 and 10 km in width, and the relative smoothness of the terminator indicates that the surface relief is fairly shallow. This mosaic shows the terminator near the south pole of the satellite.

2

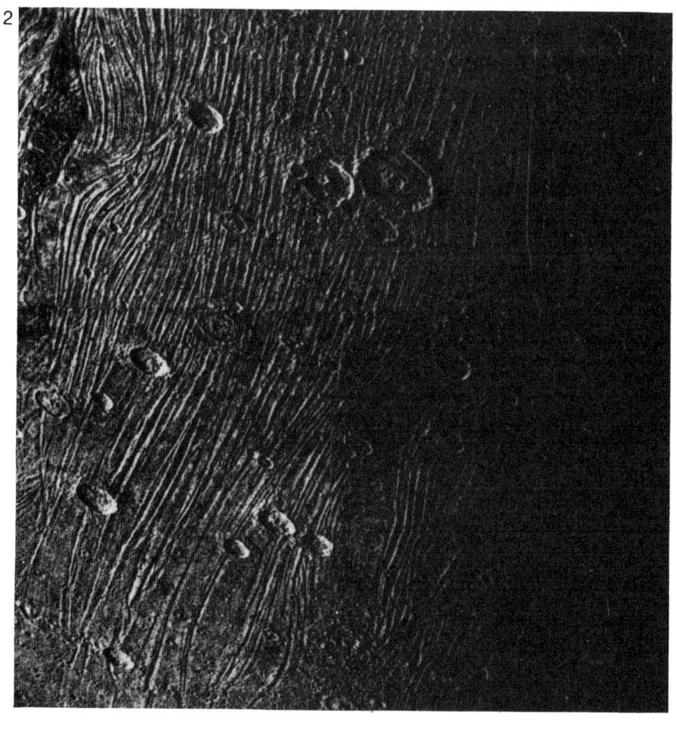

3
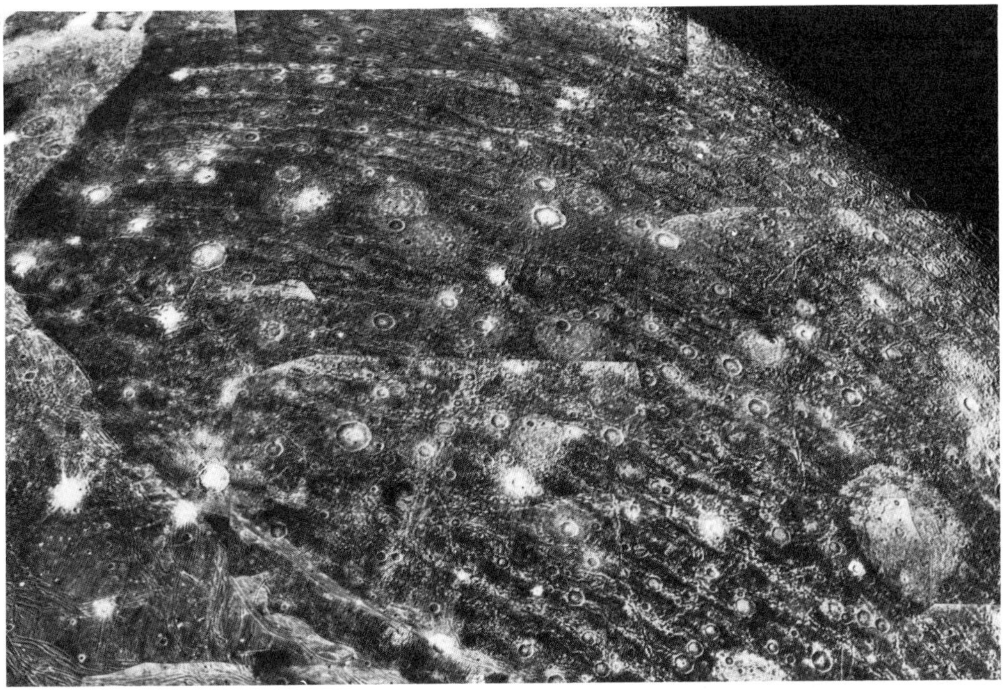

3. Galileo Regio
The most important of the large, dark types of terrain, Galileo Regio, covers about one-third of the hemisphere turned permanently away from Jupiter. It is heavily cratered, but its northern part is less dark than the rest of the interior, and may indicate some kind of condensate. Crossing Galileo Regio may be seen a series of parallel, gently curved bright streaks. These could be the result of a vast impact some distance away, but no trace of any impact center has been found, and the cause of the features may, in fact, be internal. Galileo Regio shows comparatively little vertical relief, because of the glacier-like "creep" in a crust composed largely of ice. The region is about 3,200 km in diameter, and probably represents the oldest surface on view on the satellite.

4
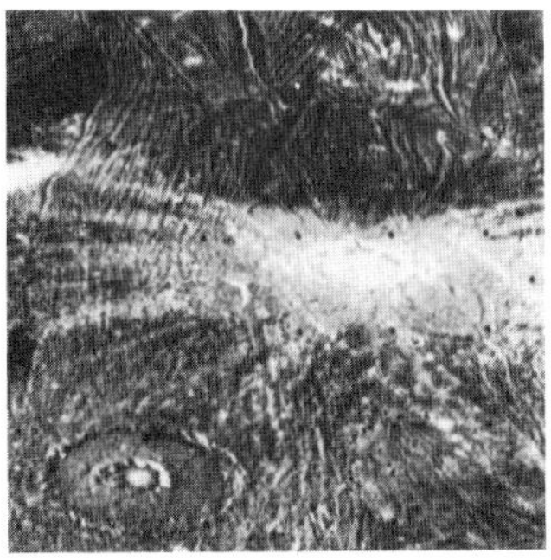

4. Ray crater (18°S, 192°W)
The ice surrounding ray craters is almost white; fresh material has been thrown up as a result of a meteoritic impact. This image shows one of the smaller unnamed ray craters; the crater at the lower left is called Eshmun.

5A

B
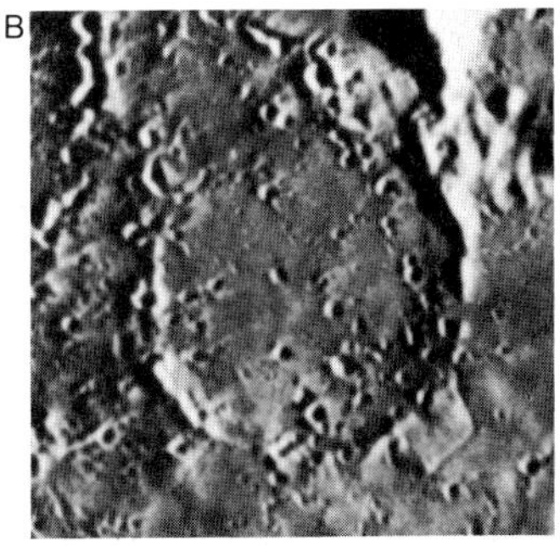

6
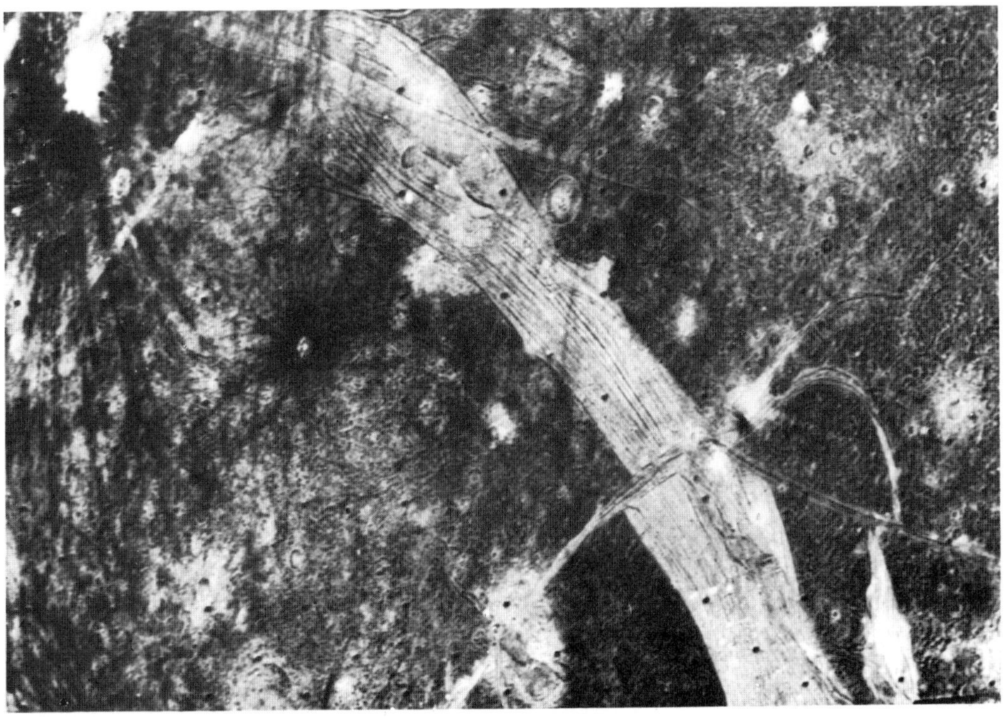

5. Ghostly crater (20°N, 120°W)
A large crater (**A**) has been almost obscured by fresher material, and a smaller, overlapping crater has been formed more recently. A photograph of the ghostly lunar crater Stadius is reproduced for comparison (**B**).

6. Tiamat Sulcus (3°S, 210°W)
A bright band of grooved terrain is shown here dividing areas of darker terrain in the Marius Regio. The bright band, Tiamat Sulcus, appears to be fractured by a fault extending from Kishar Sulcus. There is a discrepancy in the number of grooves on either side of the fault: fourteen on the northern side, as compared with twenty on the southern. The width of the grooves also differs. One explanation is that the grooves resulted from fractures that took place at different times on either side of the fault. The large number of craters suggests that the dark areas in the photograph are extremely ancient, while the brighter grooves, which exhibit fewer craters, are likely to have been formed more recently.

Ganymede Photomosaics I

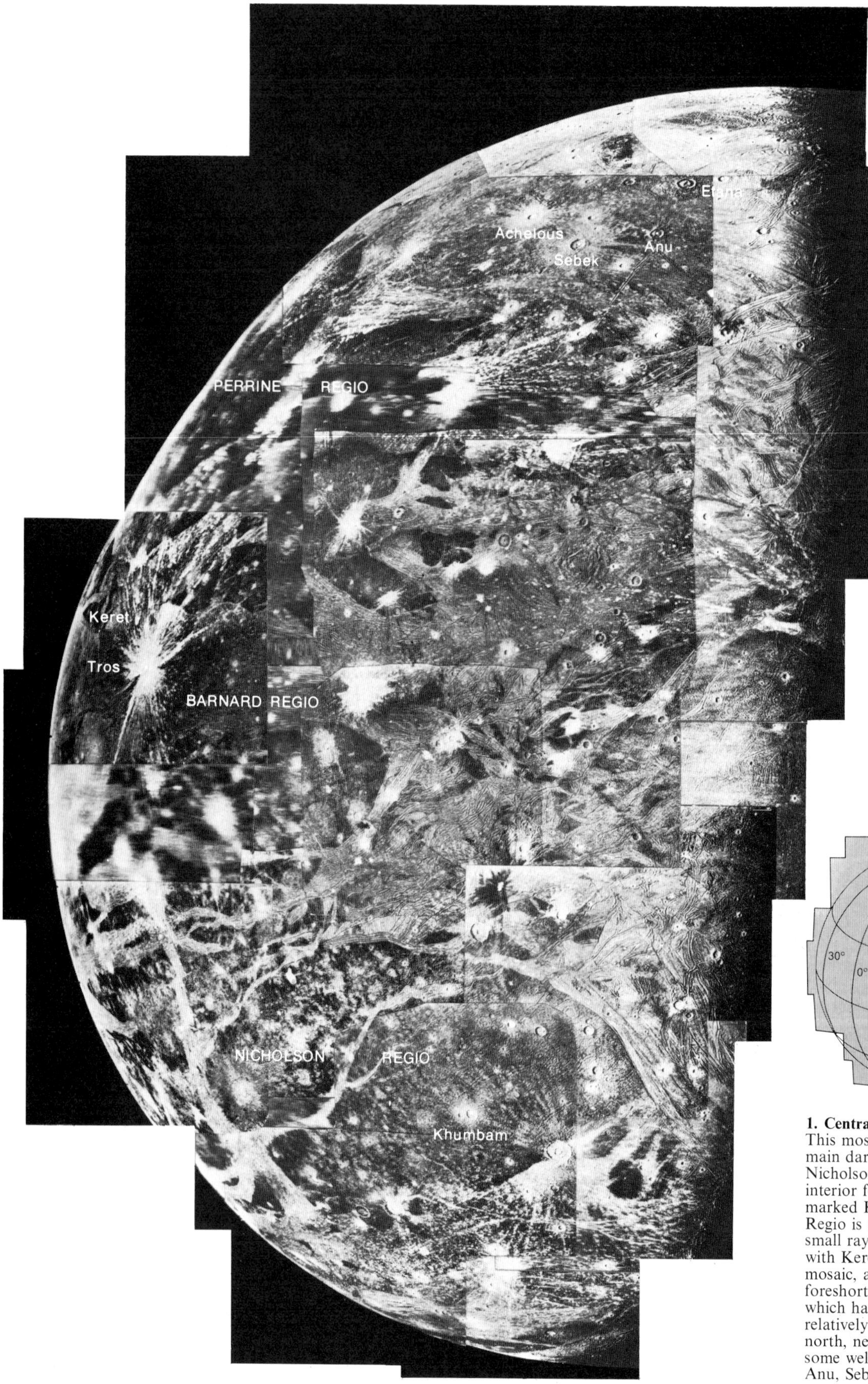

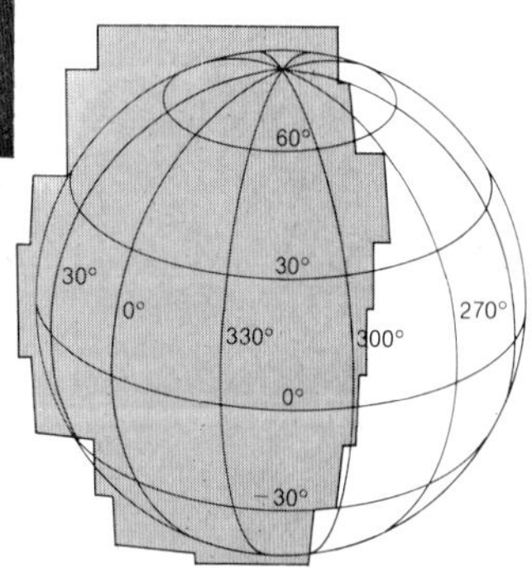

1. Central meridian
This mosaic includes one of the main dark areas on Ganymede, Nicholson Regio, with various interior features such as the well-marked Khumbam. Barnard Regio is also distinctive, with the small ray-crater Tros, together with Keret. Near the edge of the mosaic, and therefore very foreshortened, is Perrine Regio, which has a darkish interior relatively lacking in detail. The far north, near the pole, contains some well-marked craters, such as Anu, Sebek, Etana, and Achelous.

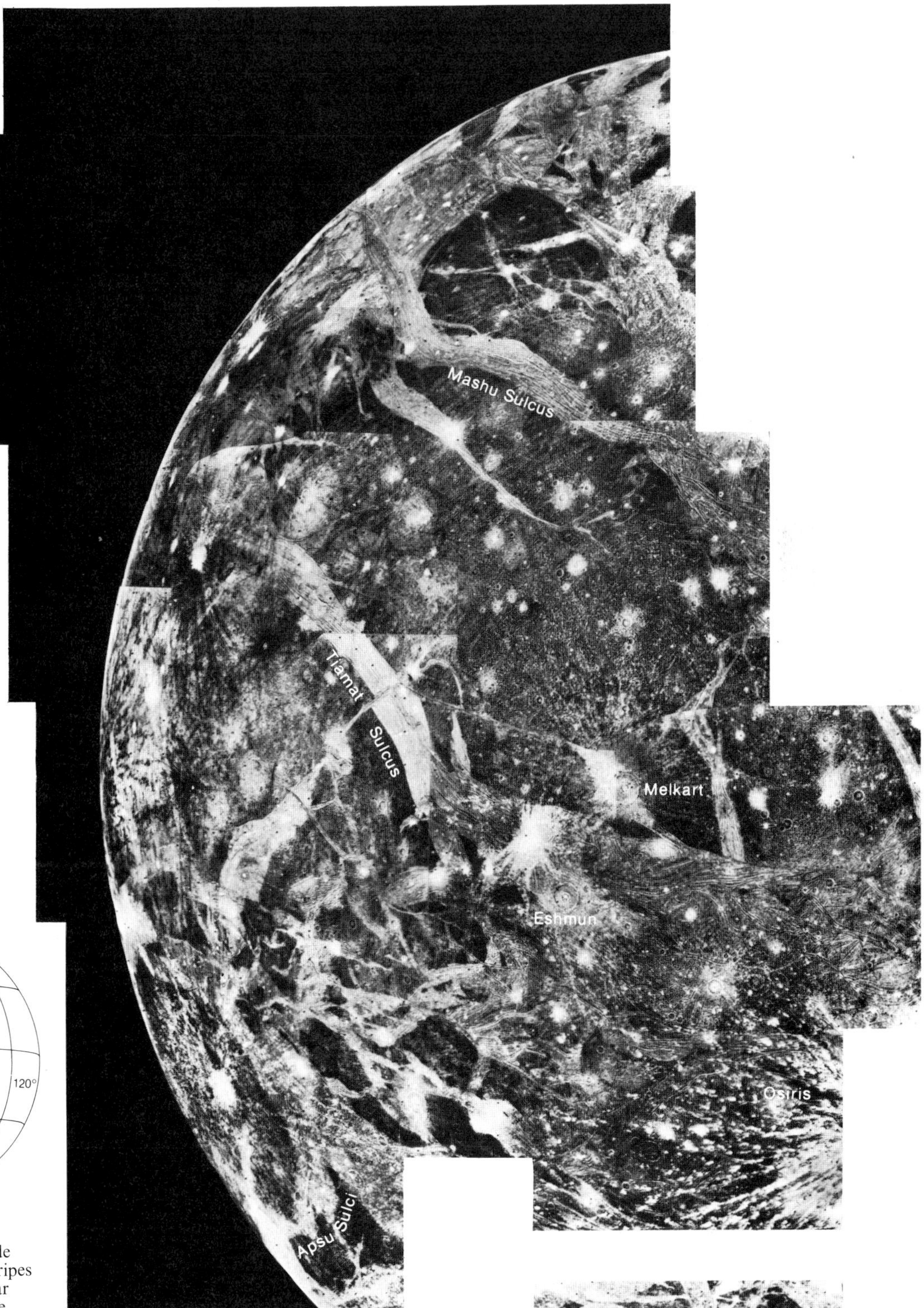

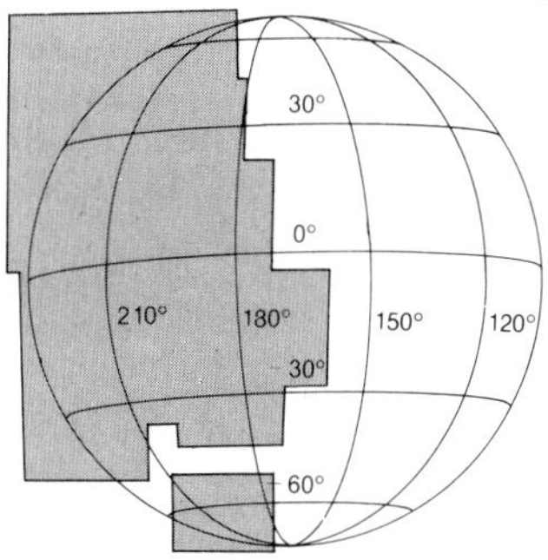

2. East region
The eastern part of Ganymede contains many of the light stripes such as Tiamat Sulcus, Kishar Sulcus and Mashu Sulcus; the overall aspect is entirely different from that of the Nicholson and Perrine areas. There are some well-marked craters here and there, notably Melkart and Eshmun, and to the lower right may be seen some of the bright streaks coming from the crater Osiris, which is just off the mosaic to the right. Apsu Sulci may be seen to the lower left.

Ganymede Photomosaics II

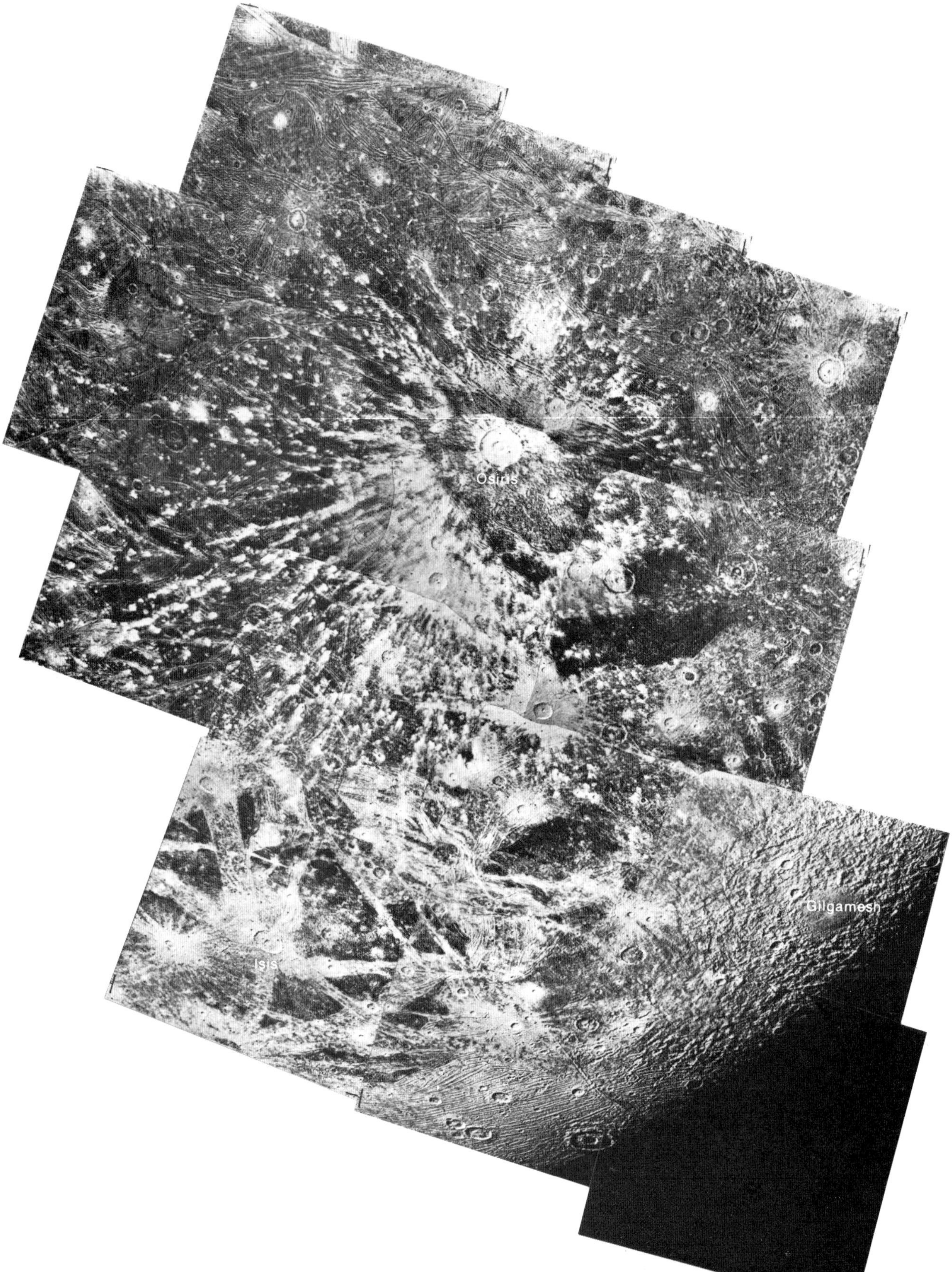

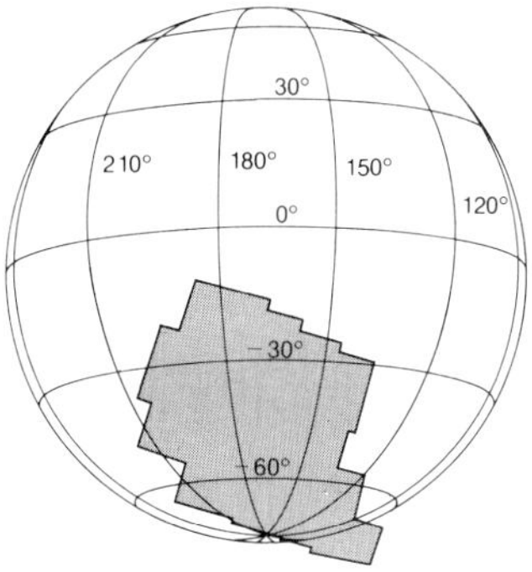

1. South polar region
This mosaic shows the southwestern region of Ganymede. It is dominated by the crater Osiris, which is bright and has a central crater inside it; it is the center of a system of bright streaks (some of which are also shown on the mosaic on page 75). Isis is another distinctive crater, in an area dominated by the strange light stripes so characteristic of Ganymede. Over to the lower right of the mosaic is a dark-floored plain, Gilgamesh, which is surrounded by an extremely rough area. There are other dark regions, as yet unnamed, and various craters, but there can be no doubt that in this part of Ganymede the most important and significant feature is Osiris.

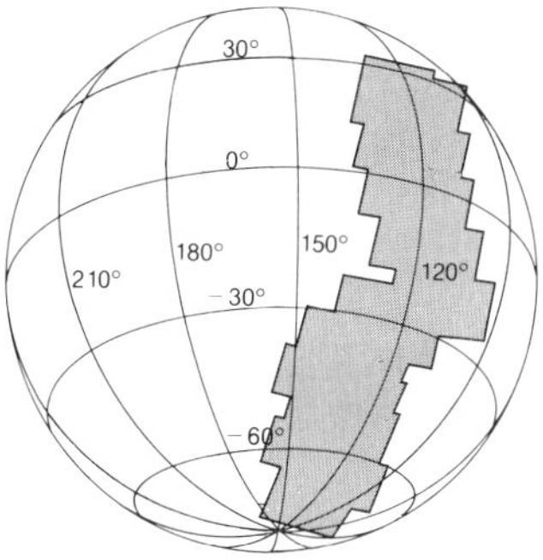

2. West region
Gilgamesh, shown on the previous mosaic, again appears here, and the extremely rough nature of the surrounding area is evident. Further north there are some well-marked craters not yet officially named, together with stripes and grooves. At the top of the mosaic (that is to say, the north) is part of the vast dark area Galileo Regio, which covers about one-third of the hemisphere turned permanently away from Jupiter; outside it to the south, separating it from the dark area below, is the Uruk Sulcus. Galileo Regio is heavily cratered, and the mosaic shows one light oval feature which is easily identified. Galileo Regio extends well beyond this mosaic.

Map of Callisto

Adal	77°N, 79°W	Dia	73°N, 56°W	Habrok	77°N, 129°W	Nerivik	22°S, 55°W
Adlinda	58°S, 20°W	Dryops	77°N, 29°W	Haki	26°N, 315°W	Nidi	66°N, 93°W
Ägröi	42°N, 12°W	Durinn	66°N, 87°W	Har	6°N, 357°W	Nori	46°N, 347°W
Akycha	74°N, 325°W	Egdir	31°N, 35°W	Hepti	64°N, 27°W	Nuada	62°N, 269°W
Alfr	9°S, 222°W	Erlik	66°N, 358°W	Hodr	69°N, 87°W	Oski	56°N, 266°W
Ali	57°N, 58°W	Fadir	56°N, 15°W	Hoenir	36°S, 261°W	Ottar	60°N, 100°W
Anarr	43°N, 3°W	Fili	65°N, 349°W	Hogni	14°S, 5°W	Pekko	17°N, 6°W
Aningan	51°N, 11°W	Finnr	14°N, 14°W	Igaluk	5°N, 315°W	Reginn	42°N, 88°W
Asgard	30°N, 140°W	Freki	82°N, 10°W	Ivarr	6°S, 322°W	Rigr	69°N, 240°W
Askr	53°N, 327°W	Frodi	69°N, 136°W	Jumo	62°N, 15°W	Sarakka	8°S, 53°W
Balkr	27°N, 12°W	Fulla	74°N, 102°W	Kari	47°N, 103°W	Seqinek	55°N, 27°W
Bavorr	48°N, 23°W	Fulnir	58°N, 37°W	Karl	56°N, 335°W	Sholmo	52°N, 18°W
Beli	61°N, 79°W	Geri	66°N, 353°W	Lodurr	52°S, 270°W	Sigyn	33°N, 27°W
Bragi	77°N, 69°W	Gipul Catena	65°N, 55°W	Loni	4°S, 215°W	Skoll	57°N, 317°W
Brami	26°N, 18°W	Gisl	56°N, 35°W	Losy	68°N, 329°W	Skuld	6°N, 37°W
Bran	25°S, 207°W	Gloi	48°N, 246°W	Mera	63°N, 73°W	Sudri	53°N, 137°W
Buga	22°N, 326°W	Goll	58°N, 323°W	Mimir	30°N, 54°W	Sumbur	69°N, 332°W
Buri	43°S, 44°W	Gondul	59°N, 115°W	Mitsina	57°N, 97°W	Tindr	5°S, 355°W
Burr	40°N, 136°W	Grimr	43°N, 214°W	Modi	67°N, 115°W	Tornarsuk	25°N, 130°W
Dag	56°N, 74°W	Gunnr	64°N, 100°W	Nama	57°N, 336°W	Tyn	68°N, 229°W
Danr	61°N, 75°W	Gymir	61°N, 55°W	Nar	4°S, 45°W	Valfodr	3°S, 246°W

Valhalla	10° N, 55° W
Vali	9° N, 327° W
Vestri	42° N, 54° W
Vitr	23° S, 347° W
Ymir	51° N, 97° W

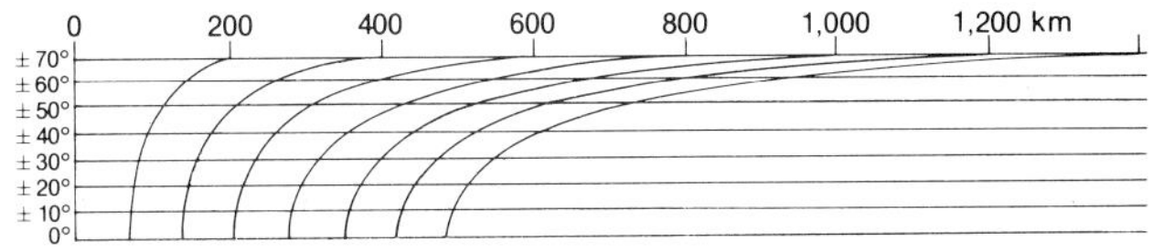

Mercator Projection

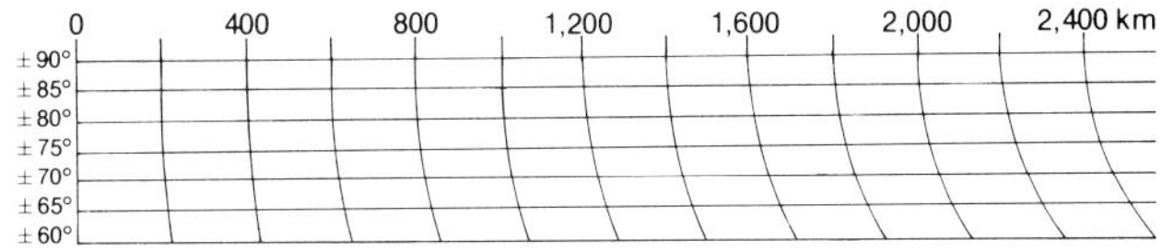

Polar Stereographic Projection

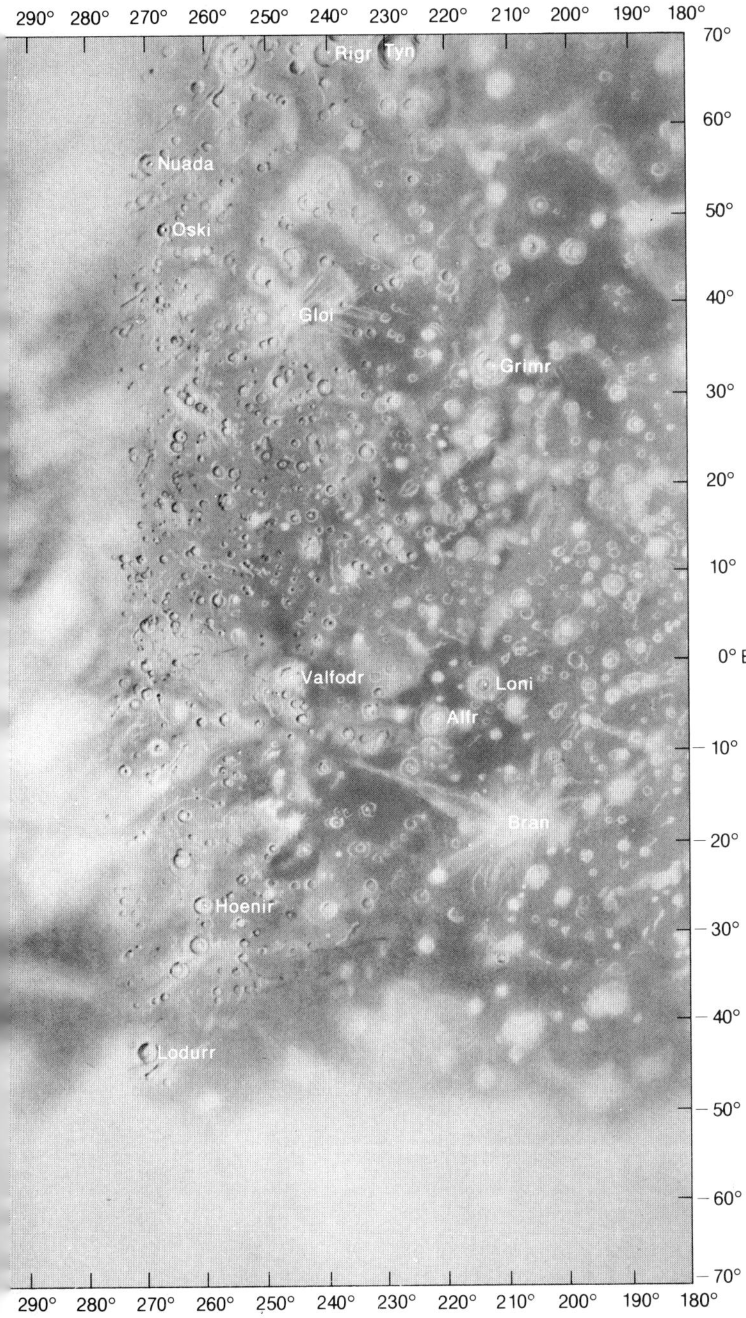

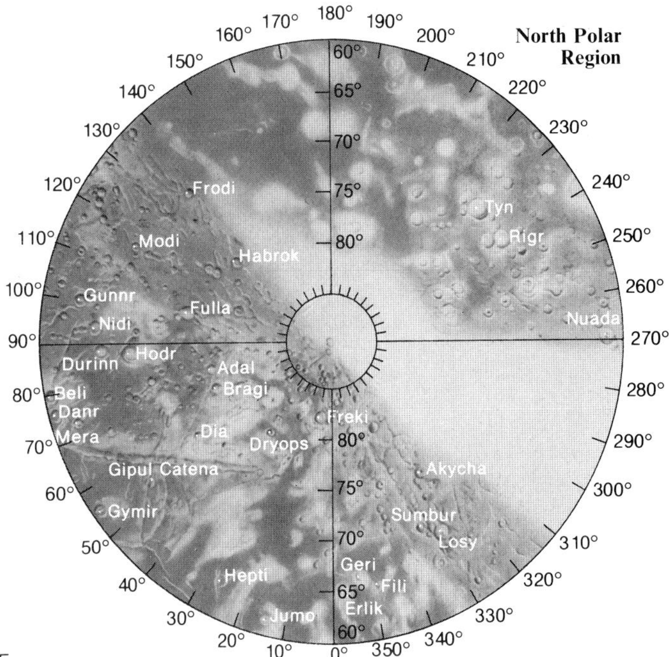

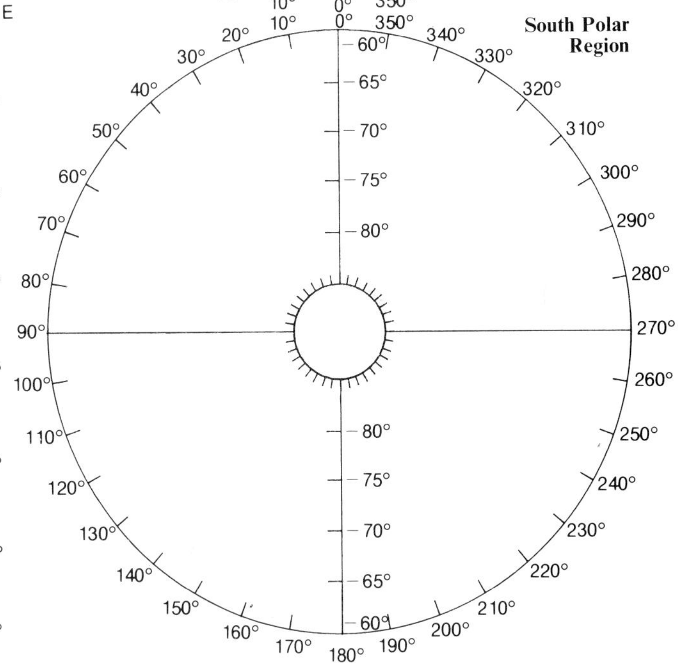

Callisto

Callisto, the outermost of the Galileans, orbits Jupiter within the extensive magnetosphere, but beyond the main radiation belts. It therefore experiences relatively little effect from the energetic particles or thermal radiation from the planet. It has a radius of about 2,500 km, and at 1.6×10^3 kg m $^{-3}$ it is the least dense of the Galileans. Indeed, this value is among the lowest of any satellite that has been measured. It is also the least reflective, with an albedo of only 0.2. Callisto therefore appears fainter than its companions, even though in size it is not much inferior to Ganymede. However, since the size and density of Callisto are known to a lower degree of accuracy than those of the other Galileans, there is a correspondingly lower degree of certainty in theoretical models of the satellite's internal structure and geological history based on these figures. Its radius, for example, is known only to within $\pm$ 150 km, as compared with $\pm$ 10 km for Ganymede.

Interior

The geology of Callisto is apparently much simpler than that of Ganymede. Following the crater-forming period, very little has happened. There is almost certainly a thick ice and rock crust, extending to a depth of up to 300 km, below which lies a convecting water or soft ice mantle which overlies a silicate core. It seems certain from the satellite's density that water accounts for a great deal of Callisto's total bulk. Tidal effects at the present time are much weaker than for Ganymede or the inner Galileans.

Surface features

Voyager 1 passed Callisto at a distance of 123,950 km and Voyager 2 at 212,510 km; as with Ganymede, some 80 percent of the surface was examined down to a resolution of 5 km or less.

The surface has been likened to "dirty ice", and it is undoubtedly very ancient. It may date back for as much as 4 thousand million years, in which case it seems to be the oldest landscape so far studied in the Solar System. Its surface is saturated with craters; Callisto is the most heavily cratered body known. Some of the craters are associated with ray systems, but the overall aspect is different from that of the Moon, because of the comparative lack of vertical relief; the terminator appears practically smooth, consistent with the picture of an icy crust. It may well be that Callisto now resembles Ganymede in its early period. There is abundant evidence of past internal activity and crustal movement upon Ganymede, but virtually none on Callisto.

There are, however, various large, concentric rings, which are still prominent even though they have been reduced by the expected crustal flow. Much the most striking is Valhalla, which lies slightly north of the satellite's equator. It is a vast structure with a brighter circular region about 600 km in diameter; the outermost of the concentric rings is almost 3,000 km across. It was the first basin to be recognized in the Jovian system, and has been compared with the Mare Orientale on the Moon, the Caloris Basin on Mercury, and Hellas on Mars, though the resemblance may well be no more than superficial. It is distinguished from these, however, by the absence of really high ridges, ring peaks or anything in the nature of a major central depression, indicating that the event which produced Valhalla caused widespread melting and flow, with the forming of shock waves in the crust, after which refreezing took place quickly enough to prevent the concentric shock rings from being destroyed. It is also notable that no ejecta patterns are visible. In the inner region there are fewer craters than elsewhere on Callisto; presumably pre-existing craters were destroyed at the time when Valhalla was formed. In the outer part of the structure, crater frequency is about the same as on the rest of the surface, but there is evidence that some of these craters are older than Valhalla itself, in which case total devastation occurred only for a radius of about 300 km around the central point. There are various other large ringed structures, notably Asgard in the south, but none to rival Valhalla.

1

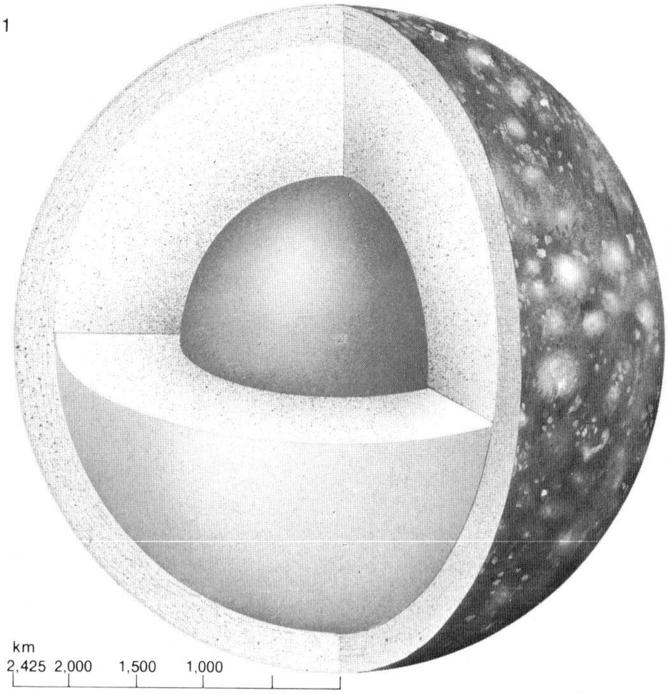

1. Interior of Callisto
Callisto is believed to have a thick crust of ice and rock extending to a depth of 200 to 300 km. Below the crust there is thought to be a 1,000 km thick mantle of convecting water or soft ice, similar to that of Ganymede; finally there is a silicate core 1,200 km in radius. Although Callisto is less reflective than the other Galileans, its surface is nevertheless thought to consist mainly of ice, its darker appearance resulting from a greater proportion of impurities.

2. Terminator
A view of the terminator near Asgard, one of the most prominent of Callisto's features, shows the surface of the satellite to be fairly smooth, at least in comparison to the Moon. However, a certain amount of relief is evident in this image as a result of Asgard's concentric rings.

2

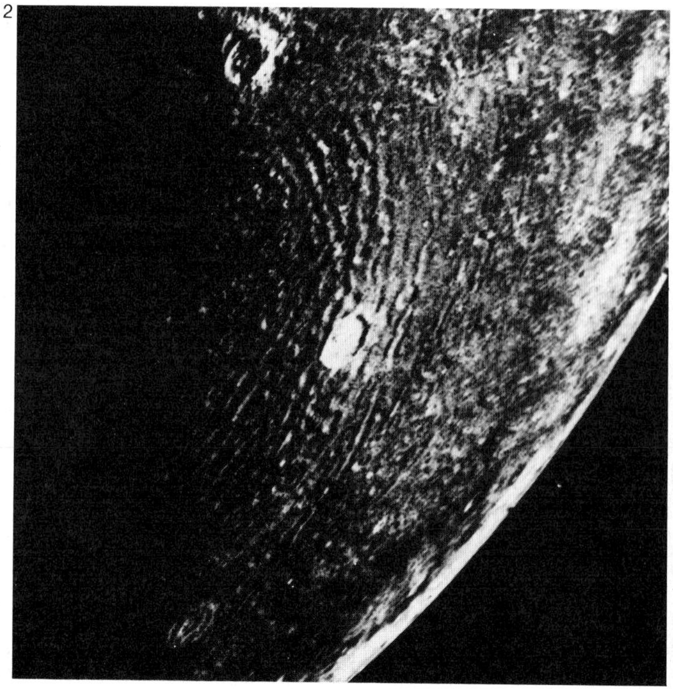

3A
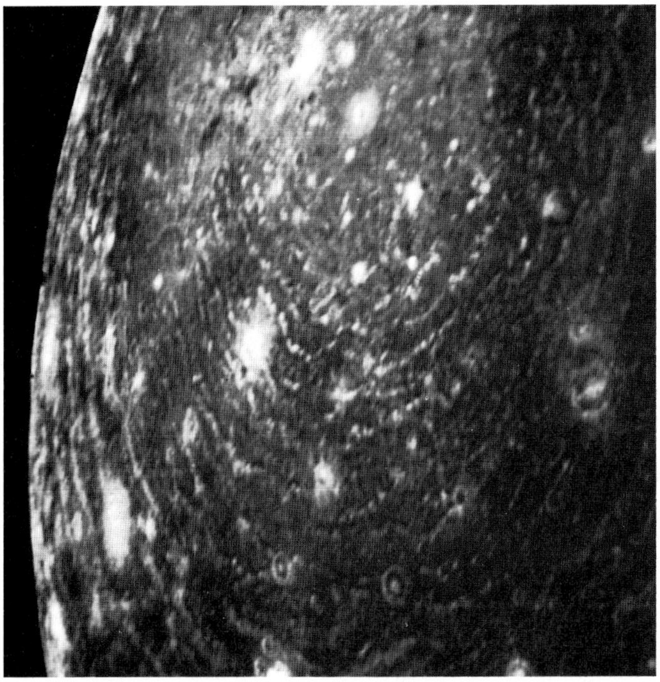

B
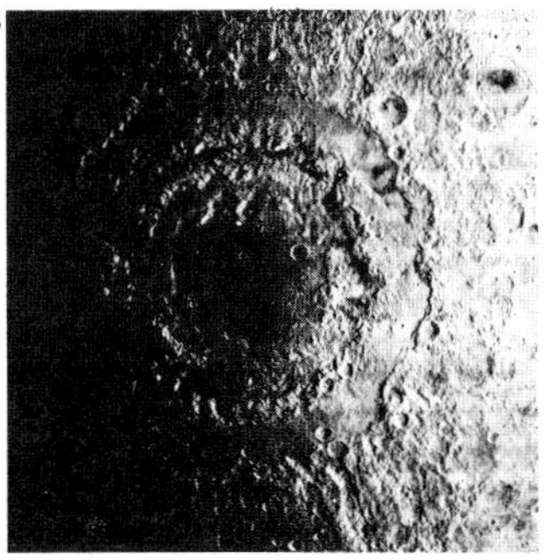

C
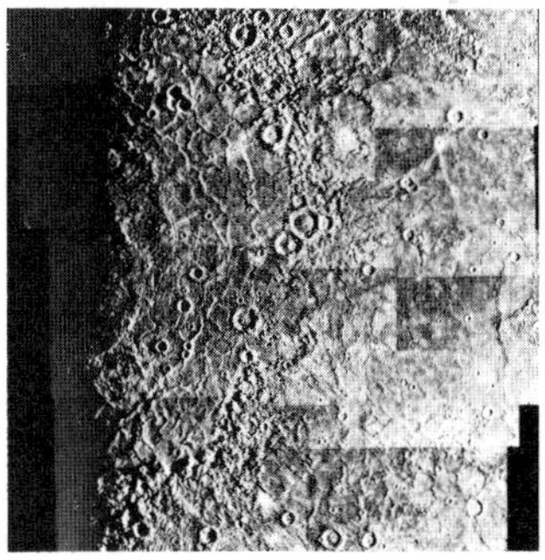

3. Valhalla (10°N, 55°W)
The largest and most striking feature on Callisto is Valhalla (**A**), a large impact basin, similar to Mare Orientale on the Moon (**B**) and Caloris Basin on Mercury (**C**). The outer ring of Valhalla is believed to have been caused by an immense impact, and it is thought that the impacting body may have come close to penetrating Callisto's crust.

4. Mosaic of Callisto
The exaggerated contrast of this image of Callisto shows up the extremely dense cratering of the satellite's surface. Crater distribution is almost uniform, but none of the craters are remarkably high. The giant feature Asgard is visible near the limb towards the upper right-hand corner of the image.

4
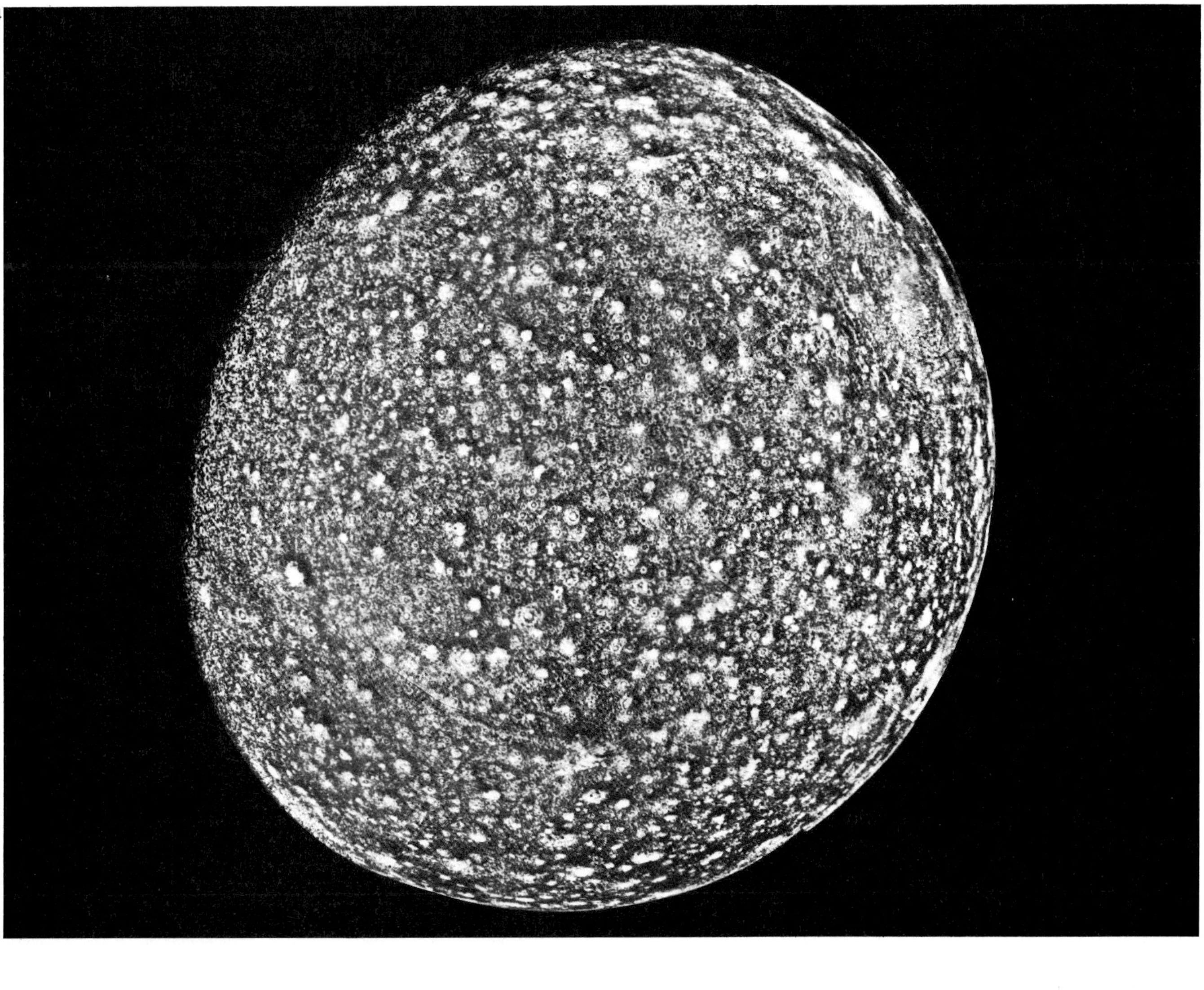

Callisto Photomosaics

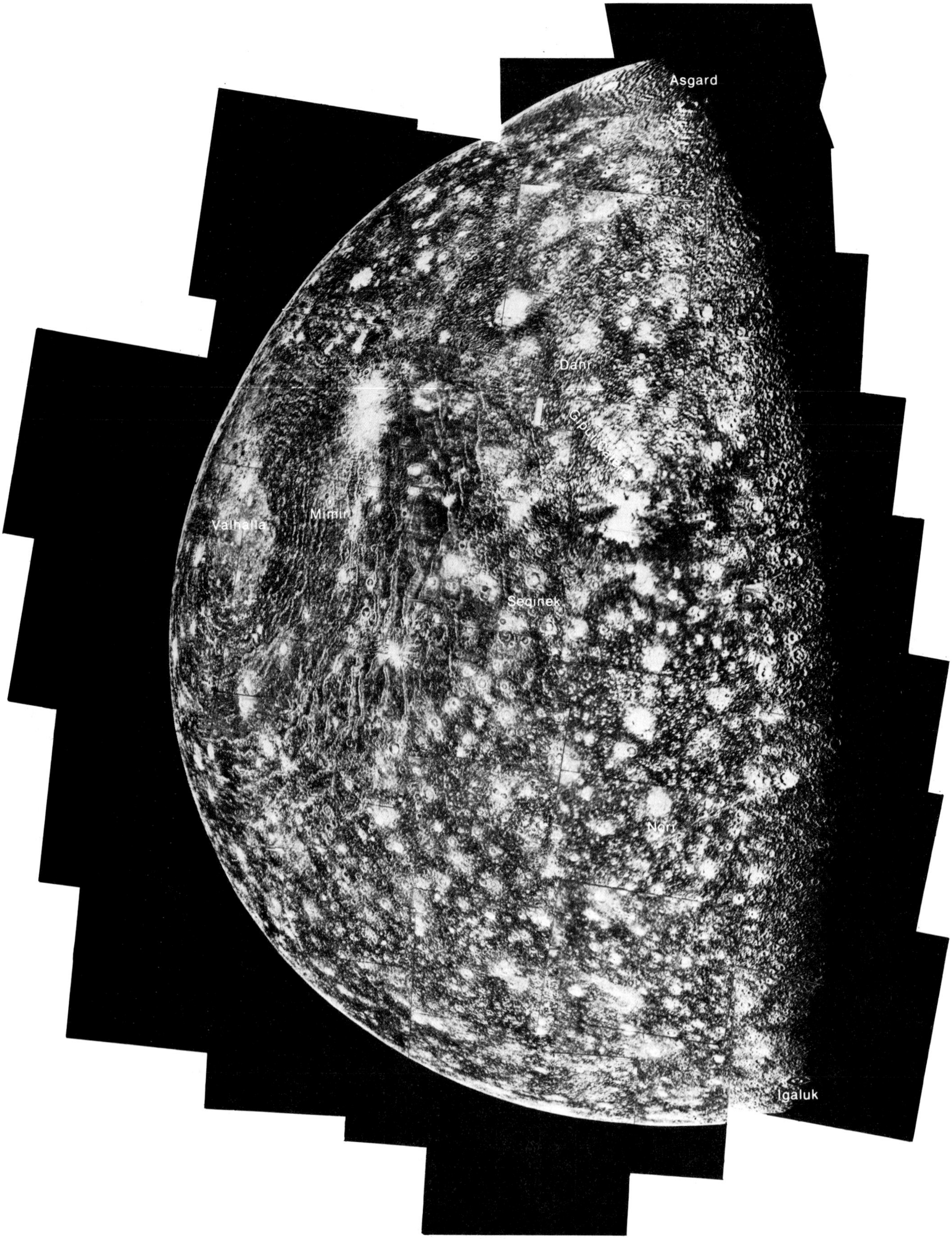

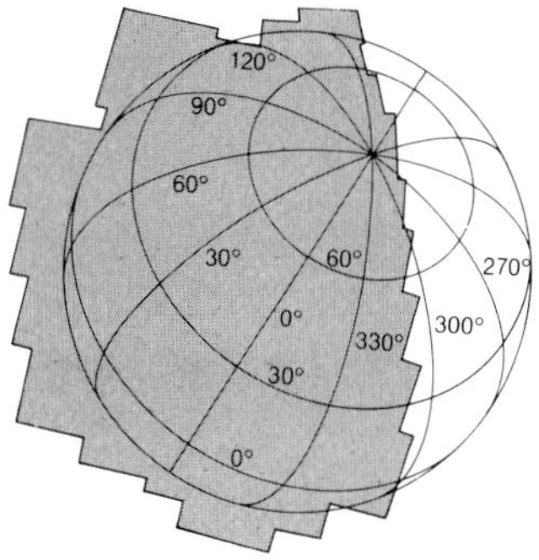

1. North region
The whole surface of Callisto is extremely ancient and heavily cratered, but there are two features of predominant importance: the ringed structures of Valhalla and Asgard. Valhalla is shown on this mosaic, to the left, and its complexity is striking. As far as Callisto is concerned, it is just as significant as Mare Orientale on the Moon or the Caloris Basin on Mercury. It is notable that there is no central structure, but there are some well-marked smaller craters such as Mimir, Skuld, Nar and Sarakka. A small part of Asgard is seen at the top of the mosaic; the north pole of Callisto is to the right, above center. There are many identifiable craters (for instance, Nori, Igaluk, Seqinek, Danr), and another feature easily recognized is Gipul Catena, a type of formation which is rather uncommon on Callisto.

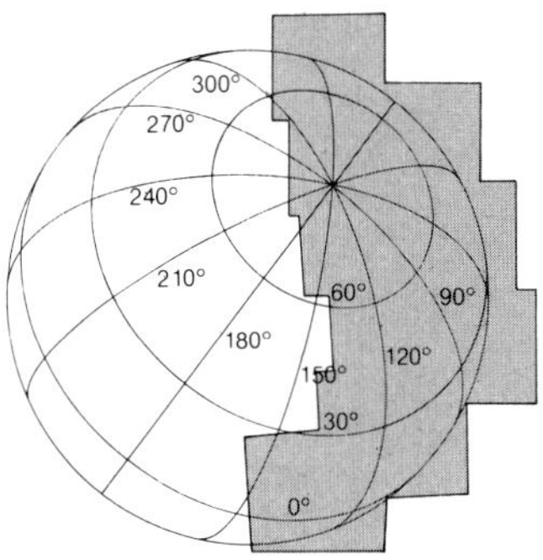

2. North-west region
The north pole of Callisto is just off the mosaic to the left, above center. Of special interest is Gipul Catena, very foreshortened on this mosaic but known to be made up of a whole chain of craters. The most important feature is the ringed basin Asgard; part of its wall is shown on the lower right. It is not nearly as large as Valhalla, but is of the same general type; on its borders are two prominent craters, Burr and Tornarsuk. The well-marked crater Sudri is also shown, above Burr and on the edge of the illuminated region. An unnamed crater with a bright nimbus is seen to the upper right of Burr and to the right of Sudri.

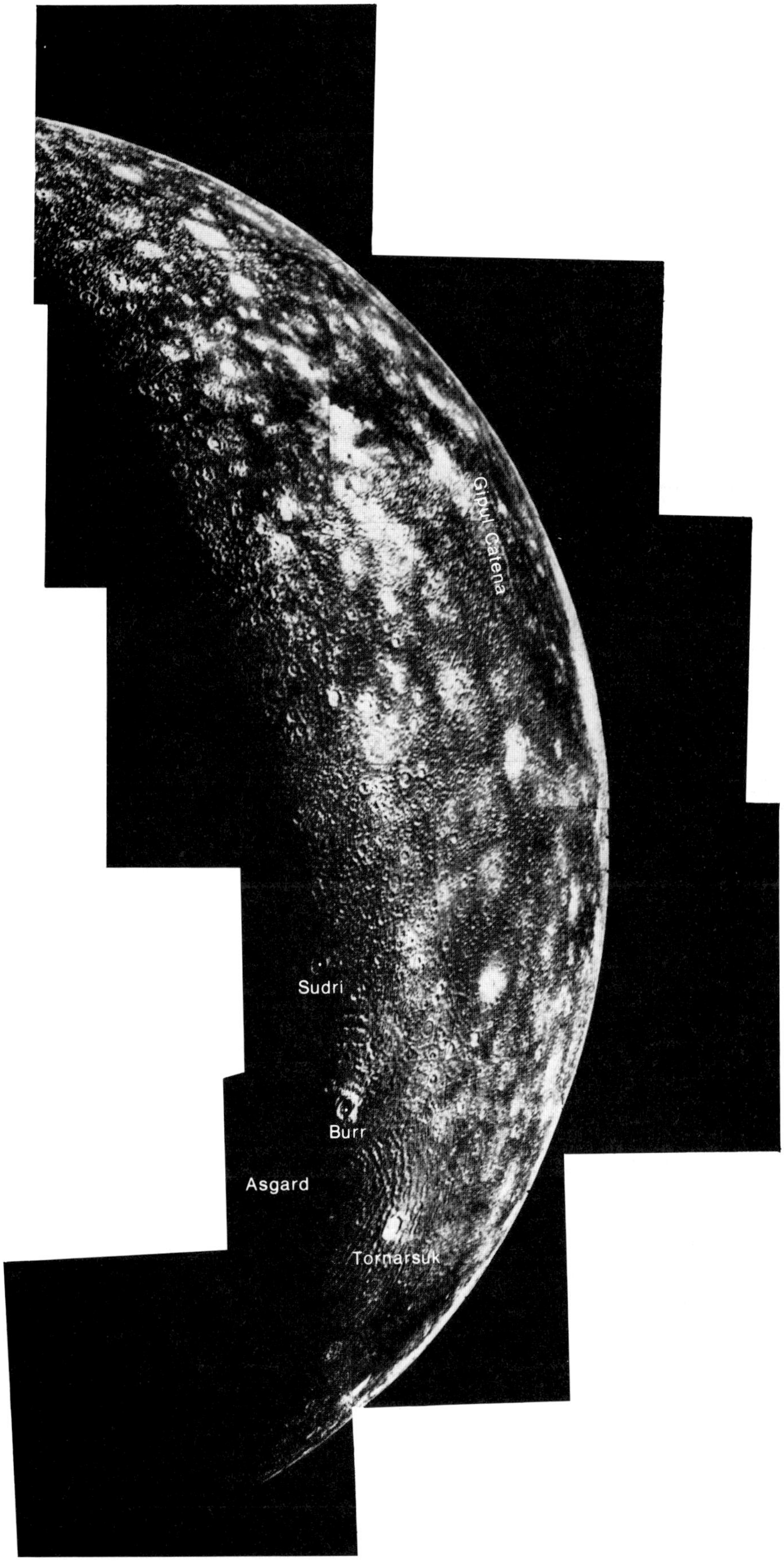

The Outer Satellites

Eight known satellites move around Jupiter beyond the orbit of Callisto. All are very faint, and only Himalia is brighter than magnitude 14. Estimates of their radii are very approximate, based on measurements of their observed brightness.

All these discoveries were photographic. For many years they remained unnamed and were referred to by number, based on the order of their discovery. However, unofficial names were used (Satellite VI, for instance, was called "Hestia"), and finally official names were ratified by the International Astronomical Union. Minor Jovian satellites with names ending with the letter "a" (Leda, Himalia, Lysithea and Elara) have direct orbital motion, while those ending in "e" (Ananke, Carme, Pasiphaë and Sinope) have retrograde motion.

Himalia and Elara

Himalia was discovered by C. D. Perrine, using the Crossley reflector at the Lick Observatory in the United States, on a plate taken on 3 November 1904. At the time Perrine was not certain whether it could be a new satellite: it might equally well have been an asteroid, but several weeks' observation sufficed to show that it was a genuine attendant of Jupiter. Early in the following year he detected an even fainter satellite, Elara. The discovery plate was taken on 2 January, again with the Crossley telescope, although it was not studied until February.

Pasiphaë and Sinope

The next discovery was made on 27 January 1908, when P. J. Melotte, at Greenwich, photographed a tiny speck of light which proved to be Satellite VIII, Pasiphaë. Meanwhile S. B. Nicholson, with the Crossley reflector, had been continuing the search, and on 21 July 1914 he discovered the ninth satellite, Sinope. Actually, the discovery was somewhat fortuitous. Nicholson had set out to photograph Pasiphaë, and had given the plate an exposure time of $2\frac{1}{2}$ hr; when he developed the plate, he found that he had recorded not only Pasiphaë but also the newcomer. Had the two satellites not been so close together, Sinope would undoubtedly have escaped detection, on this occasion at least.

Lysithea, Carme, Ananke and Leda

There matters remained for almost a quarter of a century. During the First World War the Hooker reflector at Mount Wilson in California was completed, and subsequently had a long reign as the largest and most powerful telescope in the world. With it, Nicholson made three more discoveries: Lysithea on 6 July 1938, Carme on 30 July of the same year, and Ananke on 29 September 1951. For a time the identity of Ananke was disputed, and there were suggestions that it was identical with Lysithea; a further month was needed to establish that the two were separate. Finally, in 1974, Charles Kowal at Palomar detected Leda, which is only of magnitude 20, and therefore one of the faintest known objects in the Solar System.

Orbits

Tracking these tiny bodies proved to be very much of a problem. Indeed, Pasiphaë was "lost" after its discovery in 1908, found again in 1922, lost once more until 1938, and again between 1941 and 1955. The main trouble is that the orbits alter from one revolution to another because all the satellites are so far from Jupiter that they are subject to very pronounced solar perturbations.

It is at once evident that these outer satellites fall into two main groups. Leda, Himalia, Lysithea and Elara move around Jupiter at distances of between 11 and 12 million km, while the distances of the rest range between 20 and 24 million km. Moreover, the members of the outermost group—Ananke, Carme, Pasiphaë and Sinope—have retrograde motion. Only two other retrograde satellites are known, Phoebe in Saturn's system and Triton in Neptune's, and of these only Triton is large. (Since the planet Uranus has an axial inclination of 98°—more than a right angle—and its five satellites move virtually in the equatorial plane, their movements are technically retrograde; but they are not usually reckoned as such, because they move around Uranus in the same sense as that in which Uranus rotates on its axis.)

There is no firm evidence that the outer Jovian satellites are ex-asteroids, but the retrograde motions of the more distant group are rather difficult to account for in any other way. On the other hand, the size distribution of the objects does not lend support to the idea that they are captured asteroids. An alternative suggestion is that they resulted from the break-up of larger bodies, possibly following a collision between asteroids or satellites. About their physical characteristics nothing is known at the moment, and the Voyagers did not provide any information. Himalia is the only one with a diameter of over 150 km; Elara comes next in size, but the rest are true midgets. Leda may be less than 10 km in diameter, making it even smaller than Phobos and Deimos, the tiny satellites of Mars. Seen from Jupiter, it would have an apparent diameter of less than two seconds of arc.

Although far more is known now than would have seemed possible only a year or two ago, there is clearly a great deal more to learn about the satellite system of Jupiter. There have already been many surprises, and it has been claimed that the members of the Jovian family are as interesting as Jupiter itself.

1. Leda
The photograph on which Leda was discovered was taken at Hale Observatories on 10 September 1974. Leda is the white spot indicated by an arrow.

1
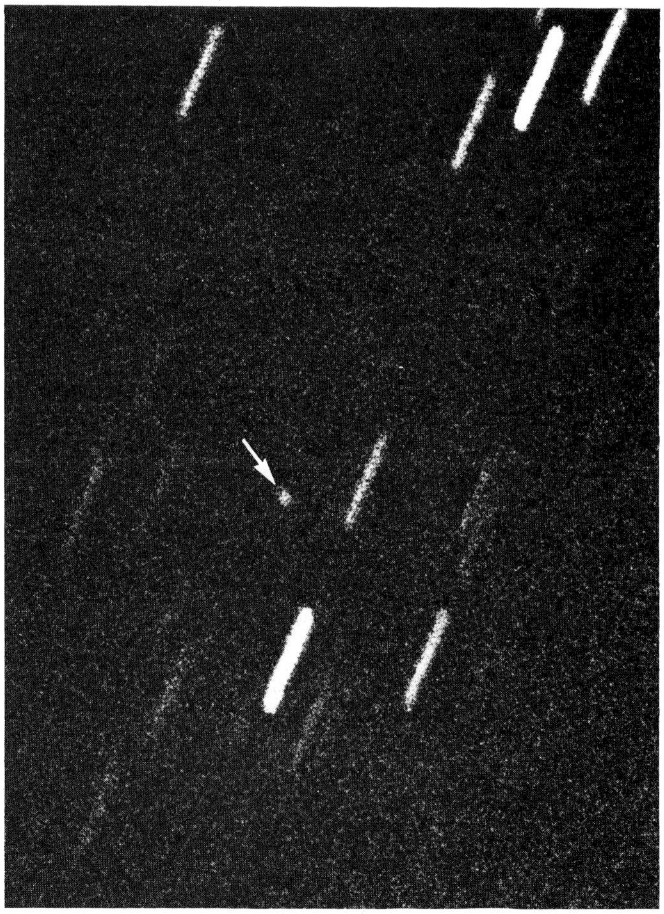

Conclusion

Any theory of the origin of the Solar System has to account for a number of known physical properties. It must explain, for example, the distribution of chemical elements throughout the System, and also the distribution of angular momentum (the product of the mass of a spinning particle, its velocity, and the distance from the center of rotation); in more general terms, the theory must explain the extreme differences between the "terrestrial" planets (Mercury, Venus, the Earth and Mars) and the major planets beyond the asteroid belt (Jupiter, Saturn, Uranus and Neptune). The former group consists of relatively small dense bodies, in contrast with the low-density objects with optically thick atmospheres which are typical of the major planets.

The "nebular hypothesis"

The most widely held modern view is known as the "nebular hypothesis", an early version of which was originally proposed in 1755 by Kant and, in a modified form, by Laplace in 1796. Neither of these early versions meets the difficulties of the problem, but more recent developments of the hypothesis succeed in accounting for at least some of the known facts. According to this theory, the Sun and stars have condensed out of clouds of interstellar dust and gas. This interstellar haze was probably thinly spread, until some sudden neighboring event, such as the explosion of a nearby star, disturbed the equilibrium. The resulting shockwave caused concentrations of particles, which, once started, continued to coalesce under the force of their own gravity. At the same time, the further it contracted, the faster the cloud would have to spin to conserve its angular momentum. However, on its own this mechanism would lead to a situation in which most of the angular momentum of the Solar System would be concentrated at the center, whereas in fact the Sun accounts for only 2 percent of the total. Jupiter, on the other hand, accounts for 60 percent. Other factors, such as radiation and magnetic forces, may have played a part in redistributing the angular momentum in such a way as to produce the results that are observed today.

The composition of material is equally important. In the absence of more detailed evidence, several alternative theories are available. According to one view, for example, the primordial nebula was originally cool, possibly containing some of the organic molecules which are essential to terrestrial life, as well as large proportions of hydrogen, helium, and other chemicals like carbon monoxide, water and ammonia. The increasing temperature of the spinning disc, however, boiled away the lighter, more volatile elements in the near vicinity of the forming star. Materials with high melting points, such as iron and the silicate materials that make up rocks, were left nearer the star, where they formed dense planets of the "terrestrial" type; the lighter elements condensed further away, so that the outer planets would be expected to contain a greater proportion of hydrogen and helium and would resemble the Sun in composition.

The origin of Jupiter

The planets are currently believed to have been formed some 4 to 5 thousand million years ago, by a process of gravitational collapse similar to the process that formed the Sun. Jupiter probably swept up more than 90 percent of the material available to the planets. At this stage, the Jovian disc of gas and dust was millions of kilometers wide, but in a few months it collapsed into a glowing red ball some 200,000 km in diameter. This ball then gradually continued to shrink to its present diameter of 143,000 km.

The initial collapse of the planet produced a large amount of heat, some of which now remains in Jupiter's interior, while the remainder has slowly leaked away into space. The residual heat has been measured directly. Without any internal source of heat, a planet at Jupiter's distance from the Sun would be expected to have a "black body" temperature of 105 K, whereas the actual value is 123 K. This excess energy is in fact consistent with a continued slow contraction of Jupiter by as little as 1 mm per year. It should be stressed, however, that this energy is totally unrelated to thermonuclear reactions; Jupiter would have to be about 100 times more massive for such reactions to take place.

The origin of Jupiter's satellites

Jupiter's extensive system of inner satellites, consisting principally of the Galileans, is believed to have been formed from the same disc of dust and gas as their parent planet; the outer satellites, on the other hand, may have a completely different origin, possibly being asteroids captured by Jupiter's gravitational field.

The densities of the inner satellites decrease with increasing distance from Jupiter, just as the densities of the planets decrease with distance from the Sun. During the formation of the satellites it is likely that heat from Jupiter prevented any water from condensing close to the planet, while it remained sufficiently cold further away, at the distance of Europa, Ganymede and Callisto, for water to condense there. As a result, a large proportion of ice and water is found on the outer Galileans, while little or none appears to be present on Io.

Craters in the Solar System

Estimating the ages of objects in the Solar System depends heavily on studies of their cratering histories. If it is correct to assume that craters are the results of collisions with meteorites and asteroids, then a great deal may be inferred about the geographical history of an area of terrain from the size, number and condition of craters on its surface. A period of intense meteoritic bombardment appears to have occurred quite early in the history of the Solar System. On the basis of the dating of the lunar samples, for example, the rate of cratering was hundreds or even thousands of times more intense 4 to 5 thousand million years ago, and has since declined, reaching its present level about 3 thousand million years ago. Mars and Mercury have provided the background to theories about the nature and origin of the impacting bodies: it turns out, for example, that the spectrum of sizes of these planetary craters matches the spectrum of sizes of the asteroids in the asteroid belt, as well as that of the meteorites that fall to Earth, strongly supporting the idea that similar bodies caused the craters.

The Galileans are important objects for the purposes of testing and developing theories of crater formation. The absence of craters on Io and their virtual absence on Europa would have been extremely damaging to the existing theories had Ganymede and Callisto been similarly devoid of craters. In the event, however, Ganymede and Callisto provided substantial confirmation of the idea that crater formation took place throughout the Solar System during an early phase in its history; eventually it may be possible to accumulate sufficient evidence to determine an absolute chronology of the Solar System. However, further analysis of rock samples from distant parts of the Solar System will be necessary before a more comprehensive theory can be developed.

Tables

The tables presented on this and the following pages are intended mainly for the practical observer of Jupiter. Specialist publications such as the *Astronomical Ephemeris* and the *American Ephemeris* appear annually and will, of course, remain the primary sources for scientific research; but it is hoped that the information provided here, devoted specifically to the planet Jupiter, brings together the most critical data in a form that will be useful as general reference or, perhaps, when more detailed works are not available.

Using the tables

Table 1 gives the dates on which Jupiter is in opposition, and thus best placed for observation, for the period 1981 to 1990 (*see* page 6). There is no opposition in 1990; the next after 1989 occurs on 28 January 1991. For much of the period covered Jupiter will be well south of the celestial equator, where it will be well placed for observers in the southern hemisphere of the Earth; it will be furthest south in 1984 when it is in the constellation of Sagittarius. The table also gives the apparent diameter and magnitude of the planet at each opposition, from which it may be seen that the largest and brightest occurs in 1987.

Tables 2 and 3 give the fundamental data of the Solar System: orbital data for the planets in Table 2, and physical data (including the Sun and Moon) in Table 3. Table 2 may be read in conjunction with diagram 1, which illustrates the elements shown in the table; however, it should be noted that the longitudes given in the table are mean rather than "true" values.

Table 4 sets out the Right Ascension and Declination of Jupiter for the period 1981–1990. Ephemeris data are not normally published more than four or five years in advance, and the accuracy of the *Astronomical Ephemeris* is only achieved for the current year, the reason for this is that considerable uncertainties exist over projections further into the future. The data in Table 1 are reproduced by courtesy of H. M. Nautical Almanac Office, who kindly supplied the computer printout from which they are taken.

Tables 5 and 6 (pages 88–89) convert the rotational periods of System I and System II to degrees of longitude moved after one-minute intervals of time: thus if the longitude of Jupiter's central meridian is known at a given moment, the tables can be used to determine its longitude at a later time. For example, on 25 June 1980 at 0 hr Universal Time, the longitude of the central meridian of Jupiter's disc is found to be 32°.5 for System I, and 47°.1 for System II. (These values are obtained from the tables published in the 1980 *Handbook of the BAA*.) Suppose an observation is made at 7 hr 14 min U.T. From Table 5, the increase in longitude of System I is 264°.6 after 7 hr 14 min, while Table 6 gives a change of 262°.3 for the same interval. Therefore at the moment of the observation the longitude of the central meridian of System I is 32°.5 + 264°.6 = 297°.1, and for System II it is 47°.1 + 262°.3 = 309°.4. (If the total comes out greater than 360°, subtract 360°.)

Tables 7 and 8 are used to determine the longitudinal drift-rate of individual features by observing them over a 30-day period. For example, suppose a white spot is observed at latitude 30° S (in other words, in System II). When it is first seen, the spot's longitude is found to be 85° W; 30 days later the same spot is 101° W: it has therefore moved +16°. From Table 8 the rotational period of the spot is found to be 9 hr 56 min 3 sec.

1A

B

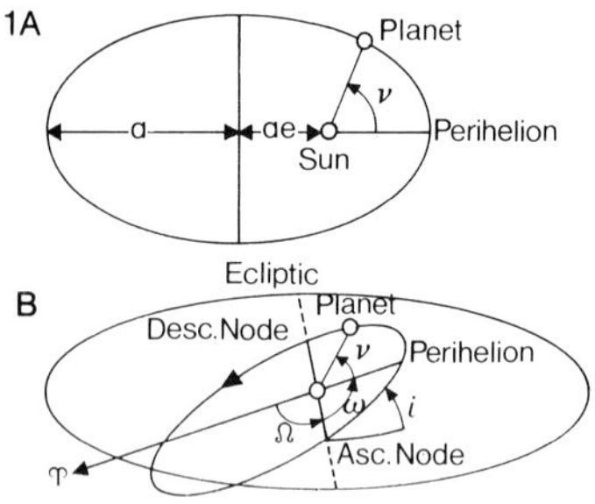

1. Orbital elements

Planets travel in elliptical orbits with the Sun at one focus. The dimensions of the ellipse may be described by two elements, the semi-major axis (a), and the eccentricity (e) as defined by diagram (**A**). Another four elements are needed to specify fully the orbit of a planet as illustrated in diagram (**B**), where the plane of the ecliptic is defined by the Earth's orbit. The angle i is the inclination of the orbit to the plane of the ecliptic; Ω is the longitude of the ascending node; ω is the "argument" of the perihelion; and L is the longitude of the planet at a specified moment. (L is given by $\Omega + \omega + \nu$.) Finally, there is the period (T).

Rotational period of Jupiter

System I: 9 hr 50 min 30.003 sec
System II: 9 hr 55 min 40.632 sec
System III: 9 hr 55 min 29.710 sec

Table 1: Oppositions 1981–1990

Date.	Apparent diam. seconds of arc	Mag.
26. 3.1981	44.2	−2.0
25. 4.1982	44.4	−2.0
27. 5.1983	45.5	−2.1
29. 6.1984	46.8	−2.2
4. 8.1985	48.5	−2.3
10. 9.1986	49.6	−2.4
18.10.1987	49.8	−2.5
23.11.1988	48.7	−2.4
27.12.1989	47.2	−2.3

Table 2: Mean elements of the planetary orbits for epoch 1980 Jan. 1.5 E.T.

Planet	Mean distance		Eccentricity	Inclination to ecliptic	of asc. node	Mean longitude of perihelion	at the epoch	Sidereal period
		a	e	i	Ω	ϖ $(=\omega+\Omega)$	L	
	A.U.	millions of km		° ′ ″	° ′ ″	° ′ ″	° ′ ″	days
Mercury	0.3870987	57.91	0.2056306	7 00 15.7	48 05 39.2	77 08 39.4	237 26 09.2	87.969
Venus	0.7233322	108.21	0.0067826	3 23 40.0	76 29 59.2	131 17 22.7	358 08 12.4	224.701
Earth	1.0000000	149.60	0.0167175	— — —	— — —	102 35 47.2	100 18 43.2	365.256
Mars	1.5236915	227.94	0.0933865	1 50 59.3	49 24 11.6	335 41 27.2	127 06 26.0	686.980
Jupiter	5.2028039	778.34	0.0484681	1 18 15.2	100 14 48.0	14 00 01.9	147 05 29.8	4332.59
Saturn	9.5388437	1,427.01	0.0556125	2 29 20.9	113 28 55.4	92 39 22.9	165 22 24.3	10,759.20
Uranus	19.181826	2,869.6	0.0472639	0 46 23.5	73 53 54.2	170 20 10.7	227 17 14.5	30,684.8
Neptune	30.058021	4,496.7	0.0085904	1 46 18.8	131 33 36.7	44 27 01.1	260 54 42.6	60,190.5
Pluto	39.44	5,900.0	0.250	17 12 00	110	223		90,465.0

Table 3: Physical data for the Sun, Moon and planets

Name	Diameter equatorial	polar	Inclination	Reciprocal mass	Mass	Density	Escape velocity	Volume	Surface gravity	Mean vis. opposition	Albedo
	km	km	degrees	(Sun = 1)	kg	(water = 1)	km s⁻¹	(Earth = 1)	(Earth = 1)	Mag.	
Sun	1,392,530	1,392,530	7.25	1	1.9891×10^{30}	1.41	617.3	1.3×10^{6}	28.0	−26.8	—
Moon	3,476	3,476	1.53	27,068,000	7.3483×10^{22}	3.34	2.37	0.02	0.165	−12.7	0.07
Mercury	4,878	4,878	0	6,023,600	3.3022×10^{23}	5.43	4.25	0.06	0.377	0.0	0.06
Venus	12,104	12,104	178	408,523.5	4.8689×10^{24}	5.24	10.36	0.86	0.902	−4.4	0.76
Earth	12,756	12,714	23.44	328,900.5	5.9742×10^{24}	5.52	11.18	1.00	1.000	—	0.36
Mars	6,794	6,759	25.20	3,098,710	6.4191×10^{23}	3.93	5.02	0.15	0.379	−2.0	0.16
Jupiter	142,800	134,200	3.12	1,047.355	1.899×10^{27}	1.32	59.6	1,323	2.69	−2.6	0.73
Saturn	120,000	108,000	26.73	3,498.5	5.684×10^{26}	0.70	35.6	752	1.19	+0.7	0.76
Uranus	52,000	49,000	97.86	22,869	8.6978×10^{25}	1.25	21.1	64	0.93	+5.5	0.93
Neptune	48,400	47,400	29.56	19,314	1.028×10^{26}	1.77	24.6	54	1.22	+7.8	0.62
Pluto	3,000	3,000	90	3,000,000	6.6×10^{23}	4.7	7.7	0.01	0.20	+14.9	0.5

Table 4: Position of Jupiter 1981–1989

Date d mo yr	R.A. h m s	Dec ° ′ ″
31. 5.80	10 17 39	+11 50 14
10. 6.80	10 21 46	+11 25 22
20. 6.80	10 26 41	+10 55 42
30. 6.80	10 32 19	+10 21 45
10. 7.80	10 38 32	+ 9 44 2
20. 7.80	10 45 15	+ 9 2 59
30. 7.80	10 52 22	+ 8 19 5
9. 8.80	10 59 49	+ 7 32 49
19. 8.80	11 7 31	+ 6 44 39
29. 8.80	11 15 23	+ 5 55 6
8. 9.80	11 23 20	+ 5 4 38
18. 9.80	11 31 20	+ 4 13 46
28. 9.80	11 39 17	+ 3 23 4
8.10.80	11 47 8	+ 2 33 5
18.10.80	11 54 47	+ 1 44 23
28.10.80	12 2 9	+ 0 57 37
7.11.80	12 9 11	+ 0 13 24
17.11.80	12 15 46	− 0 27 36
27.11.80	12 21 47	− 1 4 42
7.12.80	12 27 10	− 1 37 12
17.12.80	12 31 46	− 2 4 26
27.12.80	12 35 30	− 2 25 42
6. 1.81	12 38 14	− 2 40 27
16. 1.81	12 39 54	− 2 48 8
26. 1.81	12 40 25	− 2 48 28
5. 2.81	12 39 45	− 2 41 24
15. 2.81	12 27 57	− 2 27 11
25. 2.81	12 35 6	− 2 6 35
7. 3.81	12 31 22	− 1 40 48
17. 3.81	12 27 0	− 1 11 30
27. 3.81	12 22 19	− 0 40 43
6. 4.81	12 17 38	− 0 10 37
16. 4.81	12 13 16	+ 0 16 42
26. 4.81	12 9 32	+ 0 39 24
6. 5.81	12 6 39	+ 0 56 9
16. 5.81	12 4 45	+ 1 6 3
26. 5.81	12 3 57	+ 1 8 42
5. 6.81	12 4 14	+ 1 4 6
15. 6.81	12 5 37	+ 0 52 29
25. 6.81	12 8 0	+ 0 34 19
5. 7.81	12 11 20	+ 0 10 11
15. 7.81	12 15 32	− 0 19 18
25. 7.81	12 20 29	− 0 53 27
4. 8.81	12 26 6	− 1 31 39
14. 8.81	12 32 19	− 2 13 15
24. 8.81	12 39 1	− 2 57 35
3. 9.81	12 46 8	− 3 44 5
13. 9.81	12 53 36	− 4 32 9
23. 9.81	13 1 21	− 5 21 10
3.10.81	13 9 17	− 6 10 37
13.10.81	13 17 21	− 6 59 55
23.10.81	13 25 28	− 7 48 32
2.11.81	13 33 34	− 8 35 56
12.11.81	13 41 33	− 9 21 34
22.11.81	13 49 21	−10 4 57
2.12.81	13 56 52	−10 45 35
12.12.81	14 3 59	−11 22 58
22.12.81	14 10 37	−11 56 39
1. 1.82	14 16 37	−12 26 12
11. 1.82	14 21 54	−12 51 11
21. 1.82	14 26 19	−13 11 15
31. 1.82	14 29 45	−13 26 3
10. 2.82	14 32 5	−13 35 17
20. 2.82	14 33 16	−13 38 48
2. 3.82	14 33 13	−13 36 29
12. 3.82	14 31 56	−13 28 27
22. 3.82	14 29 31	−13 15 3
1. 4.82	14 26 4	−12 56 55
11. 4.82	14 21 50	−12 35 7
21. 4.82	14 17 6	−12 10 59
1. 5.82	14 12 10	−11 46 6
11. 5.82	14 7 24	−11 22 19
21. 5.82	14 3 7	−11 1 15
31. 5.82	13 59 35	−10 44 24
10. 6.82	13 57 0	−10 32 53
20. 6.82	13 55 29	−10 27 21
30. 6.82	13 55 6	−10 28 9
10. 7.82	13 55 51	−10 35 12
20. 7.82	13 57 43	−10 48 11
30. 7.82	14 0 38	−11 6 39
9. 8.82	14 4 30	−11 29 58
19. 8.82	14 9 15	−11 57 26
29. 8.82	14 14 49	−12 28 25
8. 9.82	14 21 4	−13 2 8
18. 9.82	14 27 57	−13 37 58
28. 9.82	14 35 22	−14 15 13
8.10.82	14 43 15	−14 53 14
18.10.82	14 51 30	−15 31 28
28.10.82	15 0 3	−16 9 10
7.11.82	15 8 49	−16 46 17
17.11.82	15 17 43	−17 21 54
27.11.82	15 26 39	−17 55 45
7.12.82	15 35 33	−18 27 26
17.12.82	15 44 17	−18 56 41
27.12.82	15 52 47	−19 23 15
6. 1.83	16 0 54	−19 46 58
16. 1.83	16 8 32	−20 7 43
26. 1.83	16 15 34	−20 25 26
5. 2.83	16 21 51	−20 40 8
15. 2.83	16 27 17	−20 51 54
25. 2.83	16 31 43	−21 0 47
7. 3.83	16 35 2	−21 6 53
17. 3.83	16 37 8	−21 10 19
27. 3.83	16 37 56	−21 11 9
6. 4.83	16 37 24	−21 9 27
16. 4.83	16 35 35	−21 5 17
26. 4.83	16 32 33	−20 58 46
6. 5.83	16 28 30	−20 50 7
16. 5.83	16 23 40	−20 39 40
26. 5.83	16 18 25	−20 27 59
5. 6.83	16 13 5	−20 15 53
15. 6.83	16 8 2	−20 4 13
25. 6.83	16 3 37	−19 54 3
5. 7.83	16 0 6	−19 46 14
15. 7.83	15 57 41	−19 41 29
25. 7.83	15 56 29	−19 40 17
4. 8.83	15 56 32	−19 42 47
14. 8.83	15 57 51	−19 48 55
24. 8.83	16 0 23	−19 58 25
3. 9.83	16 4 4	−20 10 49
13. 9.83	16 8 48	−20 25 35
23. 9.83	16 14 29	−20 42 8
3.10.83	16 21 2	−20 59 48
13.10.83	16 28 20	−21 17 59
23.10.83	16 36 19	−21 36 7
2.11.83	16 44 50	−21 53 37
12.11.83	16 53 50	−22 10 1
22.11.83	17 3 11	−22 24 53
2.12.83	17 12 47	−22 37 51
12.12.83	17 22 34	−22 48 43
22.12.83	17 32 25	−22 57 14
1. 1.84	17 42 13	−23 3 21
11. 1.84	17 51 52	−23 7 7
21. 1.84	18 1 16	−23 8 35
31. 1.84	18 10 18	−23 8 1
10. 2.84	18 18 52	−23 5 41
20. 2.84	18 26 49	−23 1 58
1. 3.84	18 34 5	−22 57 19
11. 3.84	18 40 30	−22 52 13
21. 3.84	18 45 58	−22 47 13
31. 3.84	18 50 23	−22 42 49
10. 4.84	18 53 37	−22 39 31
20. 4.84	18 55 36	−22 37 44
30. 4.84	18 56 15	−22 37 43
10. 5.84	18 55 33	−22 39 36
20. 5.84	18 53 32	−22 43 17
30. 5.84	18 50 18	−22 48 27
9. 6.84	18 46 2	−22 54 36
19. 6.84	18 40 59	−23 1 10
29. 6.84	18 35 31	−23 7 31
9. 7.84	18 30 0	−23 13 16
19. 7.84	18 24 50	−23 18 4
29. 7.84	18 20 22	−23 21 51
8. 8.84	18 16 53	−23 24 44
18. 8.84	18 14 37	−23 26 52
28. 8.84	18 13 40	−23 28 22
7. 9.84	18 14 5	−23 29 21
17. 9.84	18 15 52	−23 29 45
27. 9.84	18 18 56	−23 29 27
7.10.84	18 23 14	−23 28 12
17.10.84	18 28 37	−23 25 43
27.10.84	18 34 60	−23 21 43
6.11.84	18 42 14	−23 15 51
16.11.84	18 50 12	−23 7 51
26.11.84	18 58 47	−22 57 30
6.12.84	19 7 51	−22 44 36
16.12.84	19 17 18	−22 29 5
26.12.84	19 27 1	−22 10 55
5. 1.85	19 36 54	−21 50 11
15. 1.85	19 46 49	−21 27 5
25. 1.85	19 56 43	−21 1 50
4. 2.85	20 6 28	−20 34 49
14. 2.85	20 15 58	−20 6 26
24. 2.85	20 25 9	−19 37 10
6. 3.85	20 33 55	−19 7 38
16. 3.85	20 42 10	−18 38 25
26. 3.85	20 49 49	−18 10 13
5. 4.85	20 56 46	−17 43 46
15. 4.85	21 2 56	−17 19 48
25. 4.85	21 8 12	−16 59 4
5. 5.85	21 12 28	−16 42 20
15. 5.85	21 15 40	−16 30 15
25. 5.85	21 17 41	−16 23 26
4. 6.85	21 18 28	−16 22 16
14. 6.85	21 18 0	−16 26 53
24. 6.85	21 16 17	−16 37 9
4. 7.85	21 13 25	−16 52 29
14. 7.85	21 9 31	−17 11 52
24. 7.85	21 4 52	−17 34 0
3. 8.85	20 59 46	−17 57 11
13. 8.85	20 54 33	−18 19 47
23. 8.85	20 49 39	−18 40 16
2. 9.85	20 45 23	−18 57 18
12. 9.85	20 42 3	−19 10 1
22. 9.85	20 39 54	−19 17 53
2.10.85	20 39 2	−19 20 36
12.10.85	20 39 31	−19 18 11
22.10.85	20 41 20	−19 10 41
1.11.85	20 44 25	−18 58 15
11.11.85	20 48 39	−18 41 4
21.11.85	20 53 57	−18 19 19
1.12.85	21 0 9	−17 53 14
11.12.85	21 7 8	−17 23 2
21.12.85	21 14 47	−16 48 59
31.12.85	21 22 57	−16 11 22
10. 1.86	21 31 33	−15 30 33
20. 1.86	21 40 27	−14 46 52
30. 1.86	21 49 33	−14 0 48
9. 2.86	21 58 46	−13 12 45
19. 2.86	22 8 1	−12 23 15
1. 3.86	22 17 12	−11 32 52
11. 3.86	22 26 15	−10 42 5
21. 3.86	22 35 6	− 9 51 34
31. 3.86	22 43 40	− 9 1 55
10. 4.86	22 51 52	− 8 13 44
20. 4.86	22 59 38	− 7 27 44
30. 4.86	23 6 53	− 6 44 35
10. 5.86	23 13 33	− 6 4 57
20. 5.86	23 19 31	− 5 29 37
30. 5.86	23 24 42	− 4 59 14
9. 6.86	23 29 0	− 4 34 33
19. 6.86	23 32 19	− 4 16 16
29. 6.86	23 34 35	− 4 4 55
9. 7.86	23 35 42	− 4 1 0
19. 7.86	23 35 38	− 4 4 46
29. 7.86	23 34 22	− 4 16 4
8. 8.86	23 31 58	− 4 34 27
18. 8.86	23 28 33	− 4 58 51
28. 8.86	23 24 21	− 5 27 43
7. 9.86	23 19 37	− 5 59 4
17. 9.86	23 14 44	− 6 30 31
27. 9.86	23 10 3	− 6 59 47
7.10.86	23 5 55	− 7 24 44
17.10.86	23 2 38	− 7 43 36
27.10.86	23 0 25	− 7 55 18
6.11.86	22 59 25	− 7 59 11
16.11.86	22 59 41	− 7 55 1
26.11.86	23 1 13	− 7 43 3
6.12.86	23 3 57	− 7 23 39
16.12.86	23 7 47	− 6 57 21
26.12.86	23 12 37	− 6 24 54
5. 1.87	23 18 19	− 5 46 51
15. 1.87	23 24 46	− 5 4 1
25. 1.87	23 31 52	− 4 17 4
4. 2.87	23 39 29	− 3 26 40
14. 2.87	23 47 32	− 2 33 32
24. 2.87	23 55 55	− 1 38 21
6. 3.87	0 4 32	− 0 41 43
16. 3.87	0 13 19	+ 0 15 42
26. 3.87	0 22 12	+ 1 13 18
5. 4.87	0 31 5	+ 2 10 32
15. 4.87	0 39 56	+ 3 6 47
25. 4.87	0 48 39	+ 4 1 32
5. 5.87	0 57 10	+ 4 54 15
15. 5.87	1 5 25	+ 5 44 24
25. 5.87	1 13 19	+ 6 31 32
4. 6.87	1 20 47	+ 7 15 7
14. 6.87	1 27 43	+ 7 54 40
24. 6.87	1 34 2	+ 8 29 44
4. 7.87	1 39 36	+ 8 59 48
14. 7.87	1 44 20	+ 9 24 26
24. 7.87	1 48 7	+ 9 43 12
3. 8.87	1 50 49	+ 9 55 38
13. 8.87	1 52 21	+10 1 28
23. 8.87	1 52 38	+10 0 27
2. 9.87	1 51 40	+ 9 52 33
12. 9.87	1 49 28	+ 9 38 6
22. 9.87	1 46 9	+ 9 17 45
2.10.87	1 41 56	+ 8 52 40
12.10.87	1 37 7	+ 8 24 31
22.10.87	1 32 2	+ 7 55 21
1.11.87	1 27 6	+ 7 27 26
11.11.87	1 22 40	+ 7 3 1
21.11.87	1 19 4	+ 6 43 55
1.12.87	1 16 32	+ 6 31 40
11.12.87	1 15 14	+ 6 27 2
21.12.87	1 15 13	+ 6 30 18
31.12.87	1 16 30	+ 6 41 23
10. 1.88	1 19 0	+ 6 59 44
20. 1.88	1 22 39	+ 7 24 38
30. 1.88	1 27 20	+ 7 55 19
9. 2.88	1 32 56	+ 8 30 47
19. 2.88	1 39 21	+ 9 10 13
29. 2.88	1 46 28	+ 9 52 42
10. 3.88	1 54 10	+10 37 23
20. 3.88	2 2 21	+11 23 30
30. 3.88	2 10 57	+12 10 20
9. 4.88	2 19 52	+12 57 12
19. 4.88	2 29 2	+13 43 32
29. 4.88	2 38 21	+14 28 46
9. 5.88	2 47 45	+15 12 26
19. 5.88	2 57 10	+15 54 8
29. 5.88	3 6 31	+16 33 29
8. 6.88	3 15 43	+17 10 11
18. 6.88	3 24 41	+17 43 59
28. 6.88	3 33 19	+18 14 41
8. 7.88	3 41 31	+18 42 9
18. 7.88	3 49 12	+19 6 17
28. 7.88	3 56 14	+19 26 59
7. 8.88	4 2 30	+19 44 15
17. 8.88	4 7 51	+19 58 6
27. 8.88	4 12 11	+20 8 31
6. 9.88	4 15 21	+20 15 32
16. 9.88	4 17 14	+20 19 11
26. 9.88	4 17 45	+20 19 28
6.10.88	4 16 51	+20 16 26
16.10.88	4 14 34	+20 10 6
26.10.88	4 11 2	+20 0 41
5.11.88	4 6 27	+19 48 32
15.11.88	4 1 7	+19 34 18
25.11.88	3 55 28	+19 18 57
5.12.88	3 49 56	+19 3 40
15.12.88	3 44 55	+18 49 48
25.12.88	3 40 48	+18 38 42
4. 1.89	3 37 50	+18 31 21
14. 1.89	3 36 12	+18 28 31
24. 1.89	3 35 58	+18 30 30
3. 2.89	3 37 9	+18 37 11
13. 2.89	3 39 40	+18 48 14
23. 2.89	3 43 25	+19 3 3
5. 3.89	3 48 19	+19 20 55
15. 3.89	3 54 13	+19 41 6
25. 3.89	4 1 0	+20 2 48
4. 4.89	4 8 32	+20 25 18
14. 4.89	4 16 44	+20 47 54
24. 4.89	4 25 27	+21 10 0
4. 5.89	4 34 37	+21 31 3
14. 5.89	4 44 8	+21 50 37
24. 5.89	4 53 54	+22 8 18
3. 6.89	5 3 49	+22 23 50
13. 6.89	5 13 48	+22 37 2
23. 6.89	5 23 47	+22 47 44
3. 7.89	5 33 40	+22 55 56
13. 7.89	5 43 21	+23 1 40
23. 7.89	5 52 45	+23 5 4
2. 8.89	6 1 47	+23 6 19
12. 8.89	6 10 19	+23 5 42
22. 8.89	6 18 16	+23 3 36
1. 9.89	6 25 31	+23 0 24
11. 9.89	6 31 55	+22 56 35
21. 9.89	6 37 23	+22 52 40
1.10.89	6 41 45	+22 49 9
11.10.89	6 44 55	+22 46 32
21.10.89	6 46 46	+22 45 14
31.10.89	6 47 12	+22 45 33
10.11.89	6 46 12	+22 47 34
20.11.89	6 43 47	+22 51 10
30.11.89	6 40 4	+22 56 0
10.12.89	6 35 17	+23 1 33
20.12.89	6 29 45	+23 7 10

Tables

Table 5: Movement of the Central Meridian for System I

m	0^h	1^h	2^h	3^h	4^h	5^h	6^h	7^h	8^h	9^h	10^h	11^h
	°	°	°	°	°	°	°	°	°	°	°	°
0	0.0	36.6	73.2	109.7	146.3	182.9	219.5	256.1	292.7	329.2	5.8	42.4
1	0.6	37.2	73.8	110.4	146.9	183.5	220.1	256.7	293.3	329.8	6.4	43.0
2	1.2	37.8	74.4	111.0	147.5	184.1	220.7	257.3	293.9	330.5	7.0	43.6
3	1.8	38.4	75.0	111.6	148.2	184.7	221.3	257.9	294.5	331.1	7.6	44.2
4	2.4	39.0	75.6	112.2	148.8	185.3	221.9	258.5	295.1	331.7	8.3	44.8
5	3.0	39.6	76.2	112.8	149.4	186.0	222.5	259.1	295.7	332.3	8.9	45.4
6	3.7	40.2	76.8	113.4	150.0	186.6	223.1	259.7	296.3	332.9	9.5	46.1
7	4.3	40.8	77.4	114.0	150.6	187.2	223.8	260.3	296.9	333.5	10.1	46.7
8	4.9	41.5	78.0	114.6	151.2	187.8	224.4	260.9	297.5	334.1	10.7	47.3
9	5.5	42.1	78.6	115.2	151.8	188.4	225.0	261.6	298.1	334.7	11.3	47.9
10	6.1	42.7	79.3	115.8	152.4	189.0	225.6	262.2	298.7	335.3	11.9	48.5
11	6.7	43.3	79.9	116.5	153.0	189.6	226.2	262.8	299.4	335.9	12.5	49.1
12	7.3	43.9	80.5	117.1	153.6	190.2	226.8	263.4	300.0	336.5	13.1	49.7
13	7.9	44.5	81.1	117.7	154.3	190.8	227.4	264.0	300.6	337.2	13.7	50.3
14	8.5	45.1	81.7	118.3	154.9	191.4	228.0	264.6	301.2	337.8	14.3	50.9
15	9.1	45.7	82.3	118.9	155.5	192.1	228.6	265.2	301.8	338.4	15.0	51.5
16	9.8	46.3	82.9	119.5	156.1	192.7	229.2	265.8	302.4	339.0	15.6	52.1
17	10.4	46.9	83.5	120.1	156.7	193.3	229.9	266.4	303.0	339.6	16.2	52.8
18	11.0	47.6	84.1	120.7	157.3	193.9	230.5	267.0	303.6	340.2	16.8	53.4
19	11.6	48.2	84.7	121.3	157.9	194.5	231.1	267.7	304.2	340.8	17.4	54.0
20	12.2	48.8	85.4	121.9	158.5	195.1	231.7	268.3	304.8	341.4	18.0	54.6
21	12.8	49.4	86.0	122.5	159.1	195.7	232.3	268.9	305.5	342.0	18.6	55.2
22	13.4	50.0	86.6	123.2	159.7	196.3	232.9	269.5	306.1	342.6	19.2	55.8
23	14.0	50.6	87.2	123.8	160.3	196.9	233.5	270.1	306.7	343.3	19.8	56.4
24	14.6	51.2	87.8	124.4	161.0	197.5	234.1	270.7	307.3	343.9	20.4	57.0
25	15.2	51.8	88.4	125.0	161.6	198.1	234.7	271.3	307.9	344.5	21.1	57.6
26	15.9	52.4	89.0	125.6	162.2	198.8	235.3	271.9	308.5	345.1	21.7	58.2
27	16.5	53.0	89.6	126.2	162.8	199.4	235.9	272.5	309.1	345.7	22.3	58.9
28	17.1	53.7	90.2	126.8	163.4	200.0	236.6	273.1	309.7	346.3	22.9	59.5
29	17.7	54.3	90.8	127.4	164.0	200.6	237.2	273.7	310.3	346.9	23.5	60.1
30	18.3	54.9	91.5	128.0	164.6	201.2	237.8	274.4	310.9	347.5	24.1	60.7
31	18.9	55.5	92.1	128.6	165.2	201.8	238.4	275.0	311.6	348.1	24.7	61.3
32	19.5	56.1	92.7	129.3	165.8	202.4	239.0	275.6	312.2	348.7	25.3	61.9
33	20.1	56.7	93.3	129.9	166.4	203.0	239.6	276.2	312.8	349.4	25.9	62.5
34	20.7	57.3	93.9	130.5	167.1	203.6	240.2	276.8	313.4	350.0	26.5	63.1
35	21.3	57.9	94.5	131.1	167.7	204.2	240.8	277.4	314.0	350.6	27.2	63.7
36	21.9	58.5	95.1	131.7	168.3	204.9	241.4	278.0	314.6	351.2	27.8	64.3
37	22.6	59.1	95.7	132.3	168.9	205.5	242.0	278.6	315.2	351.8	28.4	65.0
38	23.2	59.7	96.3	132.9	169.5	206.1	242.7	279.2	315.8	352.4	29.0	65.6
39	23.8	60.4	96.9	133.5	170.1	206.7	243.3	279.8	316.4	353.0	29.6	66.2
40	24.4	61.0	97.6	134.1	170.7	207.3	243.9	280.5	317.0	353.6	30.2	66.8
41	25.0	61.6	98.2	134.7	171.3	207.9	244.5	281.1	317.6	354.2	30.8	67.4
42	25.6	62.2	98.8	135.4	171.9	208.5	245.1	281.7	318.3	354.8	31.4	68.0
43	26.2	62.8	99.4	136.0	172.5	209.1	245.7	282.3	318.9	355.4	32.0	68.6
44	26.8	63.4	100.0	136.6	173.2	209.7	246.3	282.9	319.5	356.1	32.6	69.2
45	27.4	64.0	100.6	137.2	173.8	210.3	246.9	283.5	320.1	356.7	33.2	69.8
46	28.0	64.6	101.2	137.8	174.4	211.0	247.5	284.1	320.7	357.3	33.9	70.4
47	28.7	65.2	101.8	138.4	175.0	211.6	248.1	284.7	321.3	357.9	34.5	71.0
48	29.3	65.8	102.4	139.0	175.6	212.2	248.8	285.3	321.9	358.5	35.1	71.7
49	29.9	66.5	103.0	139.6	176.2	212.8	249.4	285.9	322.5	359.1	35.7	72.3
50	30.5	67.1	103.6	140.2	176.8	213.4	250.0	286.6	323.1	359.7	36.3	72.9
51	31.1	67.7	104.3	140.8	177.4	214.0	250.6	287.2	323.7	0.3	36.9	73.5
52	31.7	68.3	104.9	141.4	178.0	214.6	251.2	287.8	324.4	0.9	37.5	74.1
53	32.3	68.9	105.5	142.1	178.6	215.2	251.8	288.4	325.0	1.5	38.1	74.7
54	32.9	69.5	106.1	142.7	179.2	215.8	252.4	289.0	325.6	2.2	38.7	75.3
55	33.5	70.1	106.7	143.3	179.9	216.4	253.0	289.6	326.2	2.8	39.3	75.9
56	34.1	70.7	107.3	143.9	180.5	217.0	253.6	290.2	326.8	3.4	40.0	76.5
57	34.8	71.3	107.9	144.5	181.1	217.7	254.2	290.8	327.4	4.0	40.6	77.1
58	35.4	71.9	108.5	145.1	181.7	218.3	254.8	291.4	328.0	4.6	41.2	77.8
59	36.0	72.6	109.1	145.7	182.3	218.9	255.5	292.0	328.6	5.2	41.8	78.4
60	36.6	73.2	109.7	146.3	182.9	219.5	256.1	292.7	329.2	5.8	42.4	79.0

Table 6: Movement of the Central Meridian for System II

m	0^h	1^h	2^h	3^h	4^h	5^h	6^h	7^h	8^h	9^h	10^h	11^h
	°	°	°	°	°	°	°	°	°	°	°	°
0	0.0	36.3	72.5	108.8	145.1	181.3	217.6	253.8	290.1	326.4	2.6	38.9
1	0.6	36.9	73.1	109.4	145.7	181.9	218.2	254.4	290.7	327.0	3.2	39.5
2	1.2	37.5	73.7	110.0	146.3	182.5	218.8	255.0	291.3	327.6	3.8	40.1
3	1.8	38.1	74.3	110.6	146.9	183.1	219.4	255.7	291.9	328.2	4.4	40.7
4	2.4	38.7	74.9	111.2	147.5	183.7	220.0	256.3	292.5	328.8	5.0	41.3
5	3.0	39.3	75.5	111.8	148.1	184.3	220.6	256.9	293.1	329.4	5.7	41.9
6	3.6	39.9	76.2	112.4	148.7	184.9	221.2	257.5	293.7	330.0	6.3	42.5
7	4.2	40.5	76.8	113.0	149.3	185.5	221.8	258.1	294.3	330.6	6.9	43.1
8	4.8	41.1	77.4	113.6	149.9	186.1	222.4	258.7	294.9	331.2	7.5	43.7
9	5.4	41.7	78.0	114.2	150.5	186.8	223.0	259.3	295.5	331.8	8.1	44.3
10	6.0	42.3	78.6	114.8	151.1	187.4	223.6	259.9	296.1	332.4	8.7	44.9
11	6.6	42.9	79.2	115.4	151.7	188.0	224.2	260.5	296.7	333.0	9.3	45.5
12	7.3	43.5	79.8	116.0	152.3	188.6	224.8	261.1	297.4	333.6	9.9	46.1
13	7.9	44.1	80.4	116.6	152.9	189.2	225.4	261.7	298.0	334.2	10.5	46.7
14	8.5	44.7	81.0	117.2	153.5	189.8	226.0	262.3	298.6	334.8	11.1	47.4
15	9.1	45.3	81.6	117.9	154.1	190.4	226.6	262.9	299.2	335.4	11.7	48.0
16	9.7	45.9	82.2	118.5	154.7	191.0	227.2	263.5	299.8	336.0	12.3	48.6
17	10.3	46.5	82.8	119.1	155.3	191.6	227.8	264.1	300.4	336.6	12.9	49.2
18	10.9	47.1	83.4	119.7	155.9	192.2	228.5	264.7	301.0	337.2	13.5	49.8
19	11.5	47.7	84.0	120.3	156.5	192.8	229.1	265.3	301.6	337.8	14.1	50.4
20	12.1	48.4	84.6	120.9	157.1	193.4	229.7	265.9	302.2	338.5	14.7	51.0
21	12.7	49.0	85.2	121.5	157.7	194.0	230.3	266.5	302.8	339.1	15.3	51.6
22	13.3	49.6	85.8	122.1	158.3	194.6	230.9	267.1	303.4	339.7	15.9	52.2
23	13.9	50.2	86.4	122.7	159.0	195.2	231.5	267.7	304.0	340.3	16.5	52.8
24	14.5	50.8	87.0	123.3	159.6	195.8	232.1	268.3	304.6	340.9	17.1	53.4
25	15.1	51.4	87.6	123.9	160.2	196.4	232.7	268.9	305.2	341.5	17.7	54.0
26	15.7	52.0	88.2	124.5	160.8	197.0	233.3	269.6	305.8	342.1	18.3	54.6
27	16.3	52.6	88.8	125.1	161.4	197.6	233.9	270.2	306.4	342.7	18.9	55.2
28	16.9	53.2	89.4	125.7	162.0	198.2	234.5	270.8	307.0	343.3	19.6	55.8
29	17.5	53.8	90.1	126.3	162.6	198.8	235.1	271.4	307.6	343.9	20.2	56.4
30	18.1	54.4	90.7	126.9	163.2	199.4	235.7	272.0	308.2	344.5	20.8	57.0
31	18.7	55.0	91.3	127.5	163.8	200.0	236.3	272.6	308.8	345.1	21.4	57.6
32	19.3	55.6	91.9	128.1	164.4	200.7	236.9	273.2	309.4	345.7	22.0	58.2
33	19.9	56.2	92.5	128.7	165.0	201.3	237.5	273.8	310.0	346.3	22.6	58.8
34	20.5	56.8	93.1	129.3	165.6	201.9	238.1	274.4	310.6	346.9	23.2	59.4
35	21.2	57.4	93.7	129.9	166.2	202.5	238.7	275.0	311.3	347.5	23.8	60.0
36	21.8	58.0	94.3	130.5	166.8	203.1	239.3	275.6	311.9	348.1	24.4	60.6
37	22.4	58.6	94.9	131.1	167.4	203.7	239.9	276.2	312.5	348.7	25.0	61.3
38	23.0	59.2	95.5	131.8	168.0	204.3	240.5	276.8	313.1	349.3	25.6	61.9
39	23.6	59.8	96.1	132.4	168.6	204.9	241.1	277.4	313.7	349.9	26.2	62.5
40	24.2	60.4	96.7	133.0	169.2	205.5	241.8	278.0	314.3	350.5	26.8	63.1
41	24.8	61.0	97.3	133.6	169.8	206.1	242.4	278.6	314.9	351.1	27.4	63.7
42	25.4	61.6	97.9	134.2	170.4	206.7	243.0	279.2	315.5	351.7	28.0	64.3
43	26.0	62.3	98.5	134.8	171.0	207.3	243.6	279.8	316.1	352.4	28.6	64.9
44	26.6	62.9	99.1	135.4	171.6	207.9	244.2	280.4	316.7	353.0	29.2	65.5
45	27.2	63.5	99.7	136.0	172.2	208.5	244.8	281.0	317.3	353.6	29.8	66.1
46	27.8	64.1	100.3	136.6	172.9	209.1	245.4	281.6	317.9	354.2	30.4	66.7
47	28.4	64.7	100.9	137.2	173.5	209.7	246.0	282.2	318.5	354.8	31.0	67.3
48	29.0	65.3	101.5	137.8	174.1	210.3	246.6	282.8	319.1	355.4	31.6	67.9
49	29.6	65.9	102.1	138.4	174.7	210.9	247.2	283.5	319.7	356.0	32.2	68.5
50	30.2	66.5	102.7	139.0	175.3	211.5	247.8	284.1	320.3	356.6	32.8	69.1
51	30.8	67.1	103.3	139.6	175.9	212.1	248.4	284.7	320.9	357.2	33.5	69.7
52	31.4	67.7	104.0	140.2	176.5	212.7	249.0	285.3	321.5	357.8	34.1	70.3
53	32.0	68.3	104.6	140.8	177.1	213.3	249.6	285.9	322.1	358.4	34.7	70.9
54	32.6	68.9	105.2	141.4	177.7	213.9	250.2	286.5	322.7	359.0	35.3	71.5
55	33.2	69.5	105.8	142.0	178.3	214.6	250.8	287.1	323.3	359.6	35.9	72.1
56	33.8	70.1	106.4	142.6	178.9	215.2	251.4	287.7	323.9	0.2	36.5	72.7
57	34.4	70.7	107.0	143.2	179.5	215.8	252.0	288.3	324.5	0.8	37.1	73.3
58	35.1	71.3	107.6	143.8	180.1	216.4	252.6	288.9	325.2	1.4	37.7	73.9
59	35.7	71.9	108.2	144.4	180.7	217.0	253.2	289.5	325.8	2.0	38.3	74.6
60	36.3	72.5	108.8	145.1	181.3	217.6	253.8	290.1	326.4	2.6	38.9	75.2

Table 7: Conversion of Change of Longitude in Thirty Days (Long.) to Rotation Period (P) for System I

Long.	P 9^h48^m	Long.	P 9^h49^m	Long.	P 9^h50^m	Long.	P 9^h51^m	Long.	P 9^h52^m
−112°.4	0^s	−67°.5	0^s	−22°.7	0^s	+21°.9	0^s	+66°.3	0^s
111.7	1	66.8	1	22.0	1	22.6	1	67.0	1
110.9	2	66.0	2	21.3	2	23.3	2	67.8	2
110.2	3	65.3	3	20.5	3	24.1	3	68.5	3
109.4	4	64.5	4	19.8	4	24.8	4	69.3	4
108.7	5	63.8	5	19.0	5	25.6	5	70.0	5
107.9	6	63.0	6	18.3	6	26.3	6	70.7	6
107.2	7	62.3	7	17.5	7	27.1	7	71.5	7
106.4	8	61.5	8	16.8	8	27.8	8	72.2	8
105.7	9	60.8	9	16.0	9	28.5	9	73.0	9
104.9	10	60.0	10	15.3	10	29.3	10	73.7	10
104.2	11	59.3	11	14.6	11	30.0	11	74.4	11
103.4	12	58.5	12	13.8	12	30.8	12	75.2	12
102.7	13	57.8	13	13.1	13	31.5	13	75.9	13
101.9	14	57.0	14	12.3	14	32.2	14	76.7	14
101.2	15	56.3	15	11.6	15	33.0	15	77.4	15
100.4	16	55.5	16	10.8	16	33.7	16	78.1	16
99.7	17	54.8	17	10.1	17	34.5	17	78.9	17
98.9	18	54.1	18	9.3	18	35.2	18	79.6	18
98.2	19	53.3	19	8.6	19	36.0	19	80.4	19
97.4	20	52.6	20	7.9	20	36.7	20	81.1	20
96.7	21	51.8	21	7.1	21	37.4	21	81.8	21
95.9	22	51.1	22	6.4	22	38.2	22	82.6	22
95.2	23	50.3	23	5.6	23	38.9	23	83.3	23
94.4	24	49.6	24	4.9	24	39.7	24	84.0	24
93.7	25	48.8	25	4.1	25	40.4	25	84.8	25
92.9	26	48.1	26	3.4	26	41.1	26	85.5	26
92.2	27	47.3	27	2.6	27	41.9	27	86.3	27
91.4	28	46.6	28	1.9	28	42.6	28	87.0	28
90.7	29	45.8	29	1.2	29	43.4	29	87.7	29
89.9	30	45.1	30	−0.4	30	44.1	30	88.5	30
89.2	31	44.4	31	+0.3	31	44.8	31	+89.2	
88.4	32	43.6	32	1.1	32	45.6	32		
87.7	33	42.9	33	1.8	33	46.3	33		
86.9	34	42.1	34	2.5	34	47.1	34		
86.2	35	41.4	35	3.3	35	47.8	35		
85.4	36	40.6	36	4.0	36	48.5	36		
84.7	37	39.9	37	4.8	37	49.3	37		
83.9	38	39.1	38	5.5	38	50.0	38		
83.2	39	38.4	39	6.3	39	50.8	39		
82.5	40	37.6	40	7.0	40	51.5	40		
81.7	41	36.9	41	7.8	41	52.2	41		
81.0	42	36.2	42	8.5	42	53.0	42		
80.2	43	35.4	43	9.2	43	53.7	43		
79.5	44	34.7	44	10.0	44	54.5	44		
78.7	45	33.9	45	10.7	45	55.2	45		
78.0	46	33.2	46	11.5	46	56.0	46		
77.2	47	32.4	47	12.2	47	56.7	47		
76.5	48	31.7	48	12.9	48	57.4	48		
75.7	49	30.9	49	13.7	49	58.2	49		
75.0	50	30.2	50	14.4	50	58.9	50		
74.2	51	29.4	51	15.2	51	59.7	51		
73.5	52	28.7	52	15.9	52	60.4	52		
72.7	53	28.0	53	16.7	53	61.1	53		
72.0	54	27.2	54	17.4	54	61.9	54		
71.2	55	26.5	55	18.1	55	62.6	55		
70.5	56	25.7	56	18.9	56	63.4	56		
69.7	57	25.0	57	19.6	57	64.1	57		
69.0	58	24.2	58	20.4	58	64.8	58		
68.2	59	23.5	59	21.1	59	65.6	59		
67.5	60	22.7	60	21.9	60	66.3	60		
−66.8		−22.0		+22.6		+67.0			

Table 8: Conversion of Change of Longitude in Thirty Days (Long.) to Rotation Period (P) for System II

Long.	P 9^h52^m	Long.	P 9^h53^m	Long.	P 9^h54^m	Long.	P 9^h55^m	Long.	P 9^h56^m	Long.	P 9^h57^m	Long.	P 9^h58^m	Long.	P 9^h59^m
−162°.6	0^s	−118°.3	0^s	−74°.1	0^s	−30°.1	0^s	+13°.7	0^s	+57.4	0^s	+101°.0	0^s	+144°.4	0^s
161.9	1	117.6	1	73.4	1	29.4	1	14.5	1	58.2	1	101.7	1	145.1	1
161.1	2	116.8	2	72.7	2	28.7	2	15.2	2	58.9	2	102.4	2	145.9	2
160.4	3	116.1	3	71.9	3	27.9	3	15.9	3	59.6	3	103.2	3	146.6	3
159.6	4	115.3	4	71.2	4	27.2	4	16.6	4	60.3	4	103.9	4	147.3	4
158.9	5	114.6	5	70.5	5	26.5	5	17.4	5	61.1	5	104.6	5	148.0	5
158.2	6	113.9	6	69.7	6	25.7	6	18.1	6	61.8	6	105.3	6	148.8	6
157.4	7	113.1	7	69.0	7	25.0	7	18.8	7	62.5	7	106.1	7	149.5	7
156.7	8	112.4	8	68.3	8	24.3	8	19.6	8	63.3	8	106.8	8	150.2	8
155.9	9	111.7	9	67.5	9	23.5	9	20.3	9	64.0	9	107.5	9	150.9	9
155.2	10	110.9	10	66.8	10	22.8	10	21.0	10	64.7	10	108.2	10	151.6	10
154.5	11	110.2	11	66.1	11	22.1	11	21.7	11	65.4	11	109.0	11	152.4	11
153.7	12	109.4	12	65.3	12	21.3	12	22.5	12	66.2	12	109.7	12	153.1	12
153.0	13	108.7	13	64.6	13	20.6	13	23.2	13	66.9	13	110.4	13	153.8	13
152.2	14	108.0	14	63.9	14	19.9	14	23.9	14	67.6	14	111.1	14	154.5	14
151.5	15	107.2	15	63.1	15	19.2	15	24.7	15	68.3	15	111.9	15	155.2	15
150.8	16	106.5	16	62.4	16	18.4	16	25.4	16	69.1	16	112.6	16	156.0	16
150.0	17	105.8	17	61.7	17	17.7	17	26.1	17	69.8	17	113.3	17	156.7	17
149.3	18	105.0	18	60.9	18	17.0	18	26.9	18	70.5	18	114.0	18	157.4	18
148.5	19	104.3	19	60.2	19	16.2	19	27.6	19	71.2	19	114.8	19	158.1	19
147.8	20	103.6	20	59.5	20	15.5	20	28.3	20	72.0	20	115.5	20	158.9	20
147.1	21	102.8	21	58.7	21	14.8	21	29.0	21	72.7	21	116.2	21	159.6	21
146.3	22	102.1	22	58.0	22	14.0	22	29.8	22	73.4	22	116.9	22	160.3	22
145.6	23	101.3	23	57.2	23	13.3	23	30.5	23	74.2	23	117.7	23	161.0	23
144.9	24	100.6	24	56.5	24	12.6	24	31.2	24	74.9	24	118.4	24	161.7	24
144.1	25	99.9	25	55.8	25	11.8	25	32.0	25	75.6	25	119.1	25	162.5	25
143.4	26	99.1	26	55.0	26	11.1	26	32.7	26	76.3	26	119.8	26	163.2	26
142.6	27	98.4	27	54.3	27	10.4	27	33.4	27	77.1	27	120.6	27	163.9	27
141.9	28	97.7	28	53.6	28	9.6	28	34.1	28	77.8	28	121.3	28	164.6	28
141.2	29	96.9	29	52.8	29	8.9	29	34.9	29	78.5	29	122.0	29	165.4	29
140.4	30	96.2	30	52.1	30	8.2	30	35.6	30	79.2	30	122.7	30	166.1	30
139.7	31	95.5	31	51.4	31	7.5	31	36.3	31	80.0	31	123.4	31	166.8	31
138.9	32	94.7	32	50.6	32	6.7	32	37.1	32	80.7	32	124.2	32	167.5	32
138.2	33	94.0	33	49.9	33	6.0	33	37.8	33	81.4	33	124.9	33	168.2	33
137.5	34	93.3	34	49.2	34	5.3	34	38.5	34	82.1	34	125.6	34	169.0	34
136.7	35	92.5	35	48.4	35	4.5	35	39.2	35	82.9	35	126.3	35	169.7	35
136.0	36	91.8	36	47.7	36	3.8	36	40.0	36	83.6	36	127.1	36	170.4	36
135.3	37	91.0	37	47.0	37	3.1	37	40.7	37	84.3	37	127.8	37	171.1	37
134.5	38	90.3	38	46.2	38	2.3	38	41.4	38	85.0	38	128.5	38	171.8	38
133.8	39	89.6	39	45.5	39	1.6	39	42.2	39	85.8	39	129.2	39	172.6	39
133.0	40	88.9	40	44.8	40	0.9	40	42.9	40	86.5	40	130.0	40	173.3	40
132.3	41	88.1	41	44.0	41	−0.1	41	43.6	41	87.2	41	130.7	41	174.0	41
131.6	42	87.4	42	43.3	42	+0.6	42	44.3	42	87.9	42	131.4	42	174.7	42
130.8	43	86.6	43	42.6	43	1.3	43	45.1	43	88.7	43	132.1	43	175.4	43
130.1	44	85.9	44	41.9	44	2.0	44	45.8	44	89.4	44	132.9	44	176.2	44
129.4	45	85.2	45	41.1	45	2.8	45	46.5	45	90.1	45	133.6	45	176.9	45
128.6	46	84.4	46	40.4	46	3.5	46	47.2	46	90.8	46	134.3	46	177.6	46
127.9	47	83.7	47	39.7	47	4.2	47	48.0	47	91.6	47	135.0	47	178.3	47
127.1	48	83.0	48	38.9	48	5.0	48	48.7	48	92.3	48	135.7	48	179.0	48
126.4	49	82.2	49	38.2	49	5.7	49	49.4	49	93.0	49	136.5	49	179.8	49
125.7	50	81.5	50	37.5	50	6.4	50	50.2	50	93.7	50	137.2	50	180.5	50
124.9	51	80.7	51	36.7	51	7.2	51	50.9	51	94.5	51	137.9	51	181.2	51
124.2	52	80.0	52	36.0	52	7.9	52	51.6	52	95.2	52	138.6	52	181.9	52
123.4	53	79.3	53	35.3	53	8.6	53	52.3	53	95.9	53	139.4	53	182.6	53
122.7	54	78.5	54	34.5	54	9.3	54	53.1	54	96.6	54	140.1	54	183.4	54
122.0	55	77.8	55	33.8	55	10.1	55	53.8	55	97.4	55	140.8	55	184.1	55
121.2	56	77.1	56	33.1	56	10.8	56	54.5	56	98.1	56	141.5	56	184.8	56
120.5	57	73.3	57	32.3	57	11.5	57	55.3	57	98.8	57	142.2	57	185.5	57
119.8	58	75.6	58	31.6	58	12.3	58	56.0	58	99.5	58	143.0	58	186.2	58
119.0	59	74.9	59	30.9	59	13.0	59	56.7	59	100.3	59	143.7	59	187.0	59
118.3	60	74.1	60	30.1	60	13.7	60	57.4	60	101.0	60	144.4	60	187.7	60
−117.6		−73.4		−29.4		+14.5		+58.2		+101.7		+145.1		+188.4	

Glossary

Albedo The ratio of the amount of light reflected by a body to the amount of light incident on it; a measure of the reflecting power of a body. A perfect reflector would have an albedo of 1. The albedos of the planets are as follows: Mercury 0.06, Venus 0.76, Earth 0.39, Mars 0.16, Jupiter 0.43, Saturn 0.61, Uranus 0.35, Neptune 0.35, Pluto 0.5.

Allotropy The property in a chemical element of existing in different forms, with distinct physical properties but capable of forming identical chemical compounds. Ozone, for example, is an allotropic form of oxygen.

Altitude In astronomy, the angular distance of a celestial body from the horizon. In conjunction with a measurement of AZIMUTH, it describes the position of an object in the sky at a given moment.

Aphelion The point or moment of greatest distance from the Sun of an orbiting body such as a planet. The opposite of PERIHELION.

Asteroid One of a large number of rocky objects, smaller than a planet but larger than a METEORITE, in orbit around the Sun. Also known as "minor planets", over 99 percent of the asteroids in the SOLAR SYSTEM lie in a belt situated between the orbits of Mars and Jupiter.

Astronomical unit A unit of distance defined by the mean distance of the Earth from the Sun and equal to 149,597,870 km.

Azimuth The angular distance along the horizon, measured in an eastward direction, between a point due north and the point at which a vertical line through a celestial object meets the horizon. (This is the normal convention for an observer in the northern hemisphere; other conventions are sometimes followed.) *See also* ALTITUDE.

Barycenter The center of gravity of a system of massive bodies; the barycenter of the Earth–Moon system, for example, lies at a point within the Earth's globe.

Black body An idealized body which reflects none of the radiation falling on it. Such a body would be a perfect absorber of radiation, and would emit a SPECTRUM determined solely by its temperature.

Bode's Law A curious numerical relationship between the distances of the various planets from the Sun.The law is often expressed in the form:

$r_n = 0.4 + 0.3 \times 2^n$,

where r_n is the distance of the planet from the Sun and n is $-\infty$ 0, 1, 2, 3 . . . in turn. The resulting values correspond surprisingly closely with the actual distances, but most astronomers consider this to be merely a coincidence.

Celestial equator The circle formed by the projection of the Earth's equator onto the surface of the CELESTIAL SPHERE.

Celestial sphere An imaginary sphere, centered on the Earth, onto whose surface the stars may be considered, for the purposes of positional measurement and calculation, to be fixed.

Chromosphere The layer of the Sun's atmosphere lying above the PHOTOSPHERE and below the CORONA.

Comet A type of heavenly body in orbit around the Sun, with several characteristics that distinguish it from the planets, satellites or asteroids. Comets typically have highly eccentric orbits, and some of them become bright objects in the sky as they approach PERIHELION, sometimes with a distinctive "tail". They are made up of a "nucleus" with a surrounding cloud of dust and gas which forms the "coma".

Conjunction The near or exact alignment of two astronomical bodies in the sky. Also used to describe an alignment between a planet and the Sun as seen from Earth. When the planet passes behind the Sun, the conjunction is called "superior"; in the special case of Mercury or Venus passing between the Sun and the Earth, the conjunction is called "inferior".

Coriolis effect The apparent deflection of a body moving in a rotating coordinate system. For example, a projectile fired northward from the Earth's equator will appear to be deflected to the east, because the point on the equator from which it is fired will be rotating faster than its target to the north. The Coriolis effect plays an important part in determining the directions of wind and ocean currents.

Corona The outermost part of the Sun's atmosphere. It is visible to the naked eye only during a total eclipse of the Sun, when it has the appearance of a halo around the Sun's obscured disc. The corona is the source of the SOLAR WIND.

Cosmic rays Extremely energetic atomic particles, principally protons, travelling through space at speeds approaching the speed of light. A proportion of cosmic rays come from the Sun, while the rest originate somewhere outside the SOLAR SYSTEM, possibly in violent events in the GALAXY.

Culmination The maximum altitude of a celestial body above the horizon.

Declination The angular distance of a celestial body from the CELESTIAL EQUATOR; one of the two celestial coordinates, roughly equivalent to latitude on the Earth, used to represent the position of a celestial object. *See also* RIGHT ASCENSION.

Doppler effect The apparent shift in the frequency of waves that occurs when there is relative motion between the source and the observer. A receding source will appear to emit waves of longer wavelength (or lower frequency) than it would if it were stationary; with an approaching source the effect is reversed, and the wavelength appears to be shorter (higher frequency).

Eclipse The partial or total disappearance of a celestial body either behind a nonluminous body or into its shadow. A solar eclipse, for example, occurs when the Sun is obscured by the Moon's disc, while a lunar eclipse takes place when the Moon passes through the cone of shadow cast by the Earth.

Ecliptic The circle on the CELESTIAL SPHERE defined by the Sun's apparent annual motion against the stellar background. The ecliptic represents the plane in which the Earth orbits the Sun and, because the Earth's rotational axis is tilted, the ecliptic is inclined to the celestial equator at an angle, known as the "obliquity of the ecliptic", which is equal to about $23\frac{1}{2}°$.

Electromagnetic radiation Radiation in the form of waves associated with electric and magnetic disturbances, which may be manifested in a variety of forms, such as light, X-rays and radio waves, depending on the wavelength. The electric and magnetic components are often represented as two waves oscillating in different planes at right angles to one another.

Elongation The angular distance of a planet from the Sun, or of a satellite from its primary planet.

Equation of time The difference between the apparent solar time and the mean time; the value of the equation of time varies throughout the year from about $-14\frac{1}{4}$ min to about $+16\frac{1}{4}$ min.

Exosphere The outermost region of the Earth's atmosphere, beyond the IONOSPHERE.

Faculae Bright patches on the PHOTOSPHERE of the Sun, normally associated with SUNSPOT groups.

First Point of Aries *See* VERNAL EQUINOX.

Flares Sudden brilliant outbursts in the outer part of the Sun's atmosphere, typically lasting only a few minutes. Generally associated with SUNSPOTS, they give rise to a type of COSMIC RAYS.

Fraunhofer lines Dark lines appearing in the spectrum of the Sun, resulting from the absorption of certain wavelengths of light by elements in the outer parts of the Sun's atmosphere.

Galaxy A large system of stars. The term "The Galaxy" refers to the particular galaxy of which the Sun is a member.

Hertzprung-Russell diagram A graph on which is plotted the LUMINOSITY of stars against their temperature or spectral type. The diagram reveals that for a given spectral type, temperature is not randomly distributed. For the most numerous group of stars (the so-called "main-sequence" stars) the higher the temperature, the brighter, in general, is the star; other groupings in the H-R diagram represent stellar types such as Red Giants and White Dwarfs which do not obey this general rule.

Ion An atom that is electrically charged as a result of having lost or gained one or more electrons.

Ionosphere The region of the Earth's atmosphere, extending from approximately 80 km to 500 km above the surface, in which radiation from the Sun ionizes a substantial proportion of air molecules. *See also* ION.

Librations Apparent oscillations of the Moon as a result of which an Earth-based observer can see the surface from a slightly different angle at different times. Over a period of time, a total of about 59 percent of the Moon's surface can be seen from Earth.

Light-year A unit of distance defined by the distance travelled by light *in vacuo* in a year, equal to 9.4607×10^{12} km or 63,240 ASTRONOMICAL UNITS. In astronomy the more commonly used unit for large distances is the PARSEC, which is equal to 3.2616 light-years.

Limb The edge of the visible disc of a celestial body.

Luminosity The total amount of energy emitted by a star per unit of time.

Magnetosphere The region around a planet within which its magnetic field predominates over the magnetic field of the surrounding interplanetary region.

Magnitude A measure of the brightness of a star or other celestial body on a numerical scale which decreases as the brightness increases. The faintest stars visible to the naked eye on a clear night are of magnitude 6; the brightest have a mean magnitude of 1. The "absolute" magnitude of a star is defined as the apparent magnitude it would have if viewed from a standard distance of 10 PARSECS.

Meridian A great circle passing through the poles either of the Earth or of the CELESTIAL SPHERE. In astronomical usage, the term usually refers to the "observer's meridian", which passes through the observer's ZENITH.

Meteor A small particle of interplanetary material that leaves a bright trail across the sky as it burns up on entering the Earth's atmosphere.

Meteorite The remains of a METEOR that reaches the surface.

Meteroid A small lump of solid meteoritic material in space.

Nodes The points at which two great circles on the CELESTIAL SPHERE intersect; in particular, the points at which the orbit of a body, such as a planet or the Moon, crosses the ECLIPTIC.

Occultation The temporary disappearance of one celestial body, usually a star, behind another, usually a planet or moon. A solar eclipse is a particular case of an occultation.

Opposition The position of a planet in its orbit when the Earth lies on a direct line between the planet and the Sun. A planet is best placed for observation when it is at opposition.

Parallax The apparent change in the position of an object due to an actual change in the position of the observer. Measurement of parallax allows the distances of distant objects to be determined.

Parsec A large unit of distance defined as the distance at which a star would have an annual PARALLAX of one second of arc, and equal to 3.0857×10^{13} km, 206,265 ASTRONOMICAL UNITS, or 3.2616 LIGHT-YEARS.

Penumbra The region of partial shadow that is formed around the region of total shadow when the source of illumination is of finite size. *See also* UMBRA. The term is also used to describe the outer part of SUNSPOTS.

Perihelion The point or moment of closest approach to the Sun of an orbiting body such as a planet. The opposite of APHELION.

Perturbations Irregularities in the orbital motion of a body due to the gravitational influence of other orbiting bodies.

Phase angle The angle defined by the position of the Sun, a body, and the Earth, measured at the body.

Photosphere The intensely luminous layer of the Sun that forms its visible surface.

Plage Bright areas of the PHOTOSPHERE associated with active areas on the Sun, caused by the presence of gas considerably hotter than its surroundings.

Planet One of the nine medium-sized bodies (including the Earth) which orbit the Sun; a similar body orbiting any other star. Unlike stars, planets do not emit their own heat or light from thermonuclear reactions in their interiors. The word "planet" is derived from a Greek word meaning "wanderer": the planets are seen to move against the background of fixed stars. An "inferior" planet is one whose orbit lies within that of the Earth, while a "superior" planet moves outside the Earth's orbit.

Polarization A special condition of ELECTROMAGNETIC RADIATION. Radiation (such as light) may be resolved into two components, one electrical, the other magnetic, at right angles to one another. When the radiation is unpolarized, the components vibrate in every direction, but if the radiation is "plane polarized", all the electrical components are arranged in planes parallel to each other, with their associated magnetic components lying at right angles to them. Other types of polarization, such as "circular" and "elliptical", are also possible.

Quadrature The position of the Moon or an outer planet when its ELONGATION is 90°.

Right ascension (R.A.) The angle, measured eastward along the CELESTIAL EQUATOR in units of hours, minutes and seconds, between the VERNAL EQUINOX and the point at which the MERIDIAN through a celestial object intersects the celestial equator. Right ascension is roughly equivalent to longitude on the Earth, and in conjunction with one other coordinate, DECLINATION, specifies the exact position of an object in the sky.

Roche limit The critical distance from the center of a planet within which gravitational forces would be insufficient to prevent a satellite from being broken up by tidal forces. For a satellite with the same density as the parent planet, the Roche limit lies at 2.4 times the radius of the planet.

Saros An interval of 6,583 days (equal to 18 years 11.3 days) after which the Sun, the Moon and the Earth return almost exactly to their previous relative positions. Consequently, the Saros period marks the interval between successive ECLIPSES of similar type and circumstance.

Sidereal period The time taken for a body to complete one orbit, as measured against the background of fixed stars. *See also* SYNODIC PERIOD.

Sidereal time A system of measurement of time based on the Earth's period of rotation, measured against the background of fixed stars. The sidereal day is taken to begin at the moment at which the VERNAL EQUINOX crosses the observer's MERIDIAN.

Solar constant The amount of energy per second that would be received in the form of solar radiation over one square meter of the Earth's surface at the Earth's mean distance from the Sun, if no radiation was absorbed by the atmosphere.

Solar cycle The periodic variation of solar activity, as manifested in the number of SUNSPOTS, the frequency of solar FLARES and various other solar phenomena. The cycle has an average period of about 11 years.

Solar System The system made up of the Sun, the planets (Mercury, Venus, Earth, Mars, Jupiter, Saturn, Uranus, Neptune and Pluto) together with their satellites, the ASTEROIDS, COMETS, METEOROIDS and interplanetary material.

Solar wind An electrically charged stream of atomic particles, mainly protons and electrons, emitted by the Sun.

Solstices The two points on the ecliptic of maximum or minimum DECLINATION; the times at which the Sun reaches these points along its annual path. The summer solstice (corresponding to the maximum declination) falls around 21 June, the winter solstice (minimum declination) around 21 December.

Spectrum The range of wavelengths or frequencies present in a sample of ELECTROMAGNETIC RADIATION. Visible radiation (i.e. light) may be resolved into its component wavelengths by passing it through a prism; white light will be spread out into a band of colors. A glowing gas under low pressure will emit radiation only at certain specific wavelengths, which appear as bright, isolated "emission" lines in its spectrum; similarly, it will only absorb radiation at these same wavelengths. When radiation is absorbed from a "continuous" spectrum, black "absorption" lines appear. *See also* FRAUNHOFER LINES.

Stratosphere The region of the Earth's atmosphere, extending from about 15 km to 50 km above the Earth's surface, between the TROPOSPHERE and the mesophere.

Sunspots Large transient patches on the PHOTOSPHERE of the Sun which appear black in contrast with the surrounding regions. The number of sunspots varies in a periodic way (*see* SOLAR CYCLE).

Synchrotron radiation Radiation emitted by electrons travelling in a strong magnetic field at speeds approaching the speed of light.

Synodic period The interval between successive CONJUNCTIONS or, more generally, between similar configurations of a celestial body, the Sun and the Earth.

Tektites Small glassy objects, found in a few restricted areas of the Earth, whose origin remains a mystery; believed to be associated with METEORITE impacts on Earth.

Terminator The boundary between the dark and the sunlit hemispheres of a planet or satellite.

Troposphere The lowest layer of the Earth's atmosphere, within which temperature decreases with increasing altitude. It extends to a height of about 15 km.

Tropopause The boundary between the TROPOSPHERE and the STRATOSPHERE.

Umbra The dark central region of a shadow. *See also* PENUMBRA.

Vernal equinox The point on the CELESTIAL SPHERE at which the ECLIPTIC crosses the CELESTIAL EQUATOR from south to north (where the direction is defined by the Sun's motion). Also known as the First Point of Aries.

Zeeman effect The splitting of spectral lines (*see* SPECTRUM) when emission or absorption occurs in the presence of a strong magnetic field.

Zenith The point on the CELESTIAL SPHERE directly above the observer.

Zodiac A belt on the CELESTIAL SPHERE extending by about 8° on either side of the ECLIPTIC, marking the region within which the Sun and the planets are always to be found. The zodiac is divided into 12 equal zones which are named after 12 constellations.

Observing Jupiter

Almost any small telescope will show the main belts on Jupiter. For serious work, a 15 cm aperture is probably the minimum requirement for a reflector. With a 20 cm reflector there is ample scope, while with a 30 cm reflector a full program of observation can be carried out. Magnifications to be used depend partly upon seeing conditions and basically upon the aperture of the telescope. It is a common mistake to try to use too high a power. A smaller, sharp image is far better than a larger, blurred one. As a rule of thumb, a telescope should bear a magnification of 20 times per centimeter of aperture.

An equatorial mounting and clock drive is not an absolute essential, except for photography, but it certainly makes observing much easier and more convenient.

Drawings

Jupiter is very obviously flattened at the poles (*see* page 6) and this must always be taken into account in drawings. It is advisable to have prepared discs printed. The exact size is not important, but should not be less than 5 cm. The phase of Jupiter as seen from Earth is so slight that it may safely be neglected.

Because of Jupiter's quick rotation, the main details should be drawn as quickly as is consistent with accuracy. The finer details can then be filled in, probably by using a higher magnification. It is wise to complete the whole sketch in fifteen minutes or less. Drawings of special features may be made at a more leisurely rate.

Measurements

Probably the most important observational program for the owner of a modest telescope is the timing of surface transits (*see* page 8). The moment when a feature is brought to the central meridian may be timed with surprising accuracy—certainly to within one minute—and by using the relevant tables (*see* pages 88–89) the longitude of the feature may then be calculated.

It may happen that no actual transit can be taken, either because of interruption by clouds or because the transit had already occurred before observing began. The experienced observer may then give an estimated time of transit. This will be of lower accuracy, and it is never wise to make estimates for features more than half an hour away from the central meridian.

For measuring belt latitudes a micrometer is essential, and the telescope must therefore be equatorial and driven. In fact, this work is of lesser importance because belt latitudes are relatively easy to measure from photographs, and in any case they do not change substantially with time.

Satellites

Only the Galileans are visible with telescopes of the size usually owned by amateurs. The main interest centers on their phenomena: eclipses, occultations, transits and shadow transits. Timings may be made, though these are now chiefly for the observer's interest rather than their scientific value. Mutual phenomena may also be observed occasionally.

Recording observations

Each observation should be accompanied by the following data: name of observer; aperture and type of telescope; magnification; time (GMT); seeing conditions (1 to 5 on the Antoniadi scale, 1 being perfect and 5 very inferior); longitude of the central meridian for Systems I and II. Observational results can be of real use only if contained in an overall program.

In Britain the national society is the British Astronomical Association, which has an energetic Jupiter Section; regular reports appear in the Association's *Journal*. In the United States similar work is correlated by the Association of Lunar and Planetary Observers, and there are equivalent organizations coordinating work in other countries.

Bibliography

General

Gehrels, T., ed., *Jupiter* (Arizona, 1976)

Peek, B. M., *The Planet Jupiter* (Faber and Faber, 1958)

The Handbook of the British Astronomical Association (1980)

Articles

Beebe, R. F. *et. al*, "Measurement of Wind Vectors, Eddy Momentum Transports, and Energy Conversions in Jupiter's Atmosphere from Voyager 1 Images", *Geophys. Res. Lett.*, **7**, 1–4 (1980)

Broadfoot, A. L. *et al*, "Extreme Ultraviolet Observations from Voyager 1 Encounter with Jupiter", *Science*, **204**, 979–982 (1979)

Ibid., "Extreme Ultraviolet Observations from Voyager 2 Encounter with Jupiter", *Science*, **206**, 962–966 (1979)

Cook II, A. F. *et al*, "First Results on Jovian Lightning", *Nature*, **280**, 794 (1979)

Eshlemann, V. R. *et al*, "Radio Science with Voyager 1 at Jupiter: Preliminary Profiles of the Atmosphere and Ionosphere", *Science*, **204**, 976–978 (1979)

Ibid., "Radio Science with Voyager at Jupiter: Initial Voyager 2 Results and a Voyager 1 Measure of the Io Torus", *Science*, **206**, 959–962 (1979)

Gore, R., "What Voyager Saw: Jupiter's Dazzling Realm", *National Geographic*, **157**, no. 1, 2–29 (1980)

Hanel, R. *et al*, "Infrared Observations of the Jovian System from Voyager 1", *Science*, **204**, 972–976 (1979)

Ibid., "Infrared Observations of the Jovian System from Voyager 2", *Science*, **206**, 952–956 (1979)

Hide, R., "Jupiter's Great Red Spot", *Scientific American*, **218**, no. 2, 74–82 (1968)

Ingersol, A. P., "The Meteorology of Jupiter", *Scientific American*, **234**, no 3, 46–56 (1976)

Ingersol, A. P. *et al*, "Zonal Velocity and Texture in the Jovian Atmosphere Inferred from Voyager Images", *Nature*, **280**, 773–775 (1979)

Jenitt, D. C. *et al*, "Discovery of a New Jupiter Satellite", *Science*, **206**, 951–952 (1979)

Maxworthy, T. *et al*, "On the Production and Interaction of Planetary Solitary Waves, Applications to the Jovian Atmosphere", *Icarus*, **33**, 388–409 (1978)

Mitchell, J. L. *et al*, "Jovian Cloud Structure and Velocity Fields", *Nature*, **280**, 776–778 (1979)

Ness, N. F. *et al*, "Magnetic Field Studies at Jupiter by Voyager 1: Preliminary Results", *Science*, **204**, 982–987 (1979)

Ibid., "Magnetic Field Studies at Jupiter by Voyager 2: Preliminary Results", *Science*, **206**, 966–972 (1979)

Owen, T. *et al*, "Jupiter's Rings", *Nature*, **281**, 442–447 (1979)

Smith, B. A. *et al*, "Jupiter through the Eyes of Voyager 1", *Science*, **204**, 951–971 (1979)

Ibid., "The Galilean Satellites and Jupiter: Voyager 2 Imaging Science Results", *Science*, **206**, 927–950 (1979).

Soderblom, L. A., "The Galilean Moons of Jupiter", *Scientific American*, **242**, no. 1, 68–83 (1980)

Stone, E. C. and Lane, A. L. "Voyager 1, Encounter with the Jovian System", *Science*, **204**, 945–948 (1979)

Terrile, R. J., "Infrared Images of Jupiter at 5-Micrometer Wavelength During the Voyager 1 Encounter", *Science*, **204**, 1,007–1,008 (1979)

Ibid., "Jupiter's Cloud Distribution Between the Voyager 1 and 2 Encounters: Results from 5-Micrometer Imaging", *Science*, **206**, 995–996 (1979)

Williams, G. P., "Planetary Circulations 1 Barotropic Representation of Jovian and Terrestrial Turbulence", *J. Atmos. Sci.*, **35**, 1,399–1,426 (1978)

Ibid., "Planetary Circulations II the Jovian Quasi-Geostrophic Regime", *J. Atmos. Sci.*, **36**, 932–968 (1979)

Index

Figures in Roman type refer to text entries; figures in *italic* refer to illustrations or captions.

Index

The publishers gratefully acknowledge the assistance of the following in the preparation of this book: R.M. Batson, P. Hedgecock, R. Terrile and J. van der Woude. Thanks are also due to Faber and Faber for permission to reproduce material from *The Planet Jupiter* by B.M. Peek.

Photographic Credits
p.5(1) G. P. Kuiper, Catalina Observatory
p.5(3) The Mansell Collection
p.10(1) Drawings from the original observational records of the Rev. T. E. R. Phillips, by courtesy of the British Astronomical Association
p.10(2) Ann Ronan Picture Library
p.13(6) Planetary Patrol photographs furnished by Lowell Observatory
p.34 Mount Lemmon Observatory/Lunar and Planetary Laboratory
p.52(2) Ann Ronan Picture Library
p.53(3, 5) Ann Ronan Picture Library
p.53(4A, 4B, 4C, 4D) A. Dollfus/Observatoire du Pic du Midi
p.54(2A, 2B) McDonald Observatory/Lunar and Planetary Laboratory
pp.58–9 U.S. Geological Survey
pp.66–7 U.S. Geological Survey
pp.70–1 U.S. Geological Survey
p.73(5B) Lunar and Planetary Laboratory
pp.78–9 U.S. Geological Survey
p.81(3B) National Space Science Data Center through the World Data Center A for Rockets and Satellites
p.84 Charles T. Kowal, California Institute of Technology

Jet Propulsion Laboratory/NASA: p.17(3); p.23(3); p.25(4A, 4B); p.28; p.29(3, 4, 5); p.30; p.33; pp.35–48; p.55(5, 6); pp. 56–7(1A, 1B); p.56(2); p.60(3); p.61(5, 6A, 6B, 7, 8); pp.62–3; pp.64–5;p.68(2); p.69(4, 5, 6); p.72(2); p.73(3, 4, 5A, 6); p.74; p.75; p.76; p.77; p.80(2); p.81(3A, 3C, 4); p.82; p.83

Illustrators
Kai Choi, Chris Forsey, Mick Saunders, Charlotte Styles